Blunt Trauma

BLUNT
TRAUMA

After the Fall of Flight 3407

With best wishes
to Ann
from Ivy Bannister

BLUNT TRAUMA

After the Fall of Flight 111

Ivy Bannister

ashfield
PRESS

First published in 2005 by

ASHFIELD PRESS • DUBLIN • IRELAND

ISBN: 1 901658 50 3

Typeset by Ashfield Press in 12.5 on 16 point Dante
Designed by
SUSAN WAINE

Printed in Ireland by
ßETAPRINT LIMITED, DUBLIN

for Richard and Andrew

Prologue

3 SEPTEMBER, 1998. 7 am, Dublin, Ireland. Groggily, I reached out and switched on the radio, which was tuned to BBC Radio 4. The first thing I heard was a headline: a transatlantic airliner had crashed into the sea off the coast of Newfoundland. Something like a wave of horrified relief washed over my sleeping conscience. I had a ticket to travel to New York City on 18 September to see my mother and sister, both of whom had apartments in Manhattan. Since the age of eleven, I had been terrified of flying, and had often woken up icy with fear during the weeks preceding a flight. Now, in a ghastly kind of way, I could rest easy. Statistically speaking, if a transatlantic jet had just been lost, well, then, wasn't mine more likely to get there safely?

The rest of the headlines passed me by, as I drifted back to sleep. The main bulletin led with the plane crash. The carrier was identified as Swissair. The flight had departed from JFK, New York, bound for Geneva. All at once I was wide-awake, my self-centred fantasy shot to hell. The flight was my sister's flight. She took it regularly, maybe five times a month. Chilled to the bone, I staggered out of bed, tumbled down the hall and hammered on the bathroom door.

Inside, my husband was shaving. He too was listening to Radio 4.

'The crash,' I blurted out. 'The Swissair crash. It's the flight Patty takes.'

Frank looked at me. 'Phone her.'

'What? Get her out of bed at two o'clock in the morning?' I couldn't bear the idea of disturbing my hard-working sister. Not if she was asleep. Besides, phone calls in the small hours are unsettling. The last time our phone had roused us had been eighteen months earlier, with news of my father's death.

Soon, I was dressed and downstairs.

'What's got you up?' my older son asked.

I told him.

'What are the chances that she was on it?' he asked.

'One in five. Maybe one in six.'

Richard nodded. At eighteen years of age, he was calm and sensible. Quite reasonably, he suggested that the odds were in his aunt's favour. She was probably OK. Soon he was out the door, headed for his summer job as a computer games tester. A few minutes later, Frank left for work too. Our younger child, Andrew, who was nine, sensed that something was wrong. 'What's the matter? What are you worried about?'

'Nothing,' I said. 'I hope it's nothing.'

'But what *is* it?' he asked sweetly.

'Only something that *might* have happened. I'll tell you when I find out.' Andrew was happy with that, so I left him eating breakfast. Up in my study, I took out the telephone directory. I scribbled down the Dublin numbers for Delta (the airline that employed Patty) and Swissair (the airline that she flew for on a code-share agreement). My plan was to telephone both, as soon as I'd left Andrew to school.

By 8 am, we were doing his violin practise together, as we did every morning. A recent week at a Suzuki violin camp had fired Andrew's enthusiasm – so much so – that he was learning four pieces simultaneously: three lively works by Seitz and the first movement of Vivaldi's A Minor Concerto. The bright, joyous music filled the living room, but I was in turmoil.

Ten years earlier, I had worried when a Pan Am plane had exploded over Scotland, at Lockerbie, killing all on board and some on the ground. I knew then that my sister only occasionally worked on that particular flight. This time it was different. The Geneva flight was the only flight that Patty took. She had been flying the route exclusively for a couple of years. She loved both it and Swissair. The Swiss penchant for order, discipline and predictability agreed with her. And the Geneva flight was filled with successful people – bankers, diplomats and businessmen – people whom Patty admired, regular passengers she had come to know.

Andrew worked his way through his practise. Outside, I heard the neighbours' children leave for school. We played a few more bars, then I told Andrew to put away his violin. In the ensuing silence, I heard the answering machine, upstairs in my husband's study. It was the signal you get at the end of a message, after someone has just hung up. I had been so caught up in the music and my malaise that I hadn't heard the telephone.

Before I got to the top of the stairs, it was ringing again. This time, it was Frank. Having just arrived at his office, he had logged on to the Swissair website. There was a press release about the crash, but no information about the passengers or crew. There was, however, a Geneva phone number, which I jotted down. I hung up, then pressed the playback button on the flashing answering machine. My mother's voice, distant but calm, began to fill my head. 'In case you haven't heard already,' the voice said, 'your sister was on Flight 111. We have lost her.' Then the message was over.

'Oh Jesus, Jesus,' I heard somebody cry, an appalling wail. For a moment I listened, wondering who was making that terrible noise. Then I realised that it was me. Downstairs, Andrew, waiting for school, had begun to play the piano. He was playing from the score of *Titanic*, a movie that fascinated him. He had chosen the lament, the mournful music that unravels near the end, as a lifeboat rows through countless frozen corpses, bobbing in the sea.

With shaking hands, I rang my husband back. 'It was my mother on the answering machine. Patty was on the airplane.'

Patty

Patty, my only sibling, was dead. I knew it as a certainty. When a jet crashes into the sea from cruising altitude, there are no survivors. On the news it had said that wreckage had been found, but no survivors. Besides, it had been six hours since the crash. So even if the impossible had happened, who could live for six hours after a crash into the icy ocean?

'Come on,' I said to Andrew. 'Let's go to school.'

As we marched along the road, Andrew chattered contentedly about whom he was – or wasn't – going to invite to his birthday party. I didn't say much. Andrew absorbed my anxiety. 'What's the matter?' he demanded. 'I want to know.' There was no reason not to talk about it now. Anyway, from his questions it seemed that he had worked it out anyway.

'Did the plane crash?'

'It did.'

'Where did it crash?'

'Into the Atlantic Ocean. Off the coast of Newfoundland.'

'Is Auntie Patty dead?'

'Yes.'

'She might not be,' he said hopefully.

I thought about the movies that Andrew loved, where heroes and heroines were saved from peril in the most extreme circumstances. In the movies, the good guys can even be brought back from the dead, like in *Superman 2*, when Superman spins the world backwards to undo Lois Lane's death, or when Han Solo is defrosted in *The Return of the Jedi*. 'Real life isn't like that,' I had told Andrew, dozens of times. I had even shown him a picture of Christopher Reeve in his wheelchair, the Superman actor who became a paraplegic after a riding accident.

'Andrew, Auntie Patty is dead.'

'But she took me to see *Home Alone 3*,' he said persuasively. 'And can't Auntie Patty swim?'

'No one could survive in the cold water.'

We walked along in silence. We were coming to the point where I usually left him to make the final approach to the school on his own.

'I might cry in class,' Andrew said.

'Would you like me to tell your teacher what happened?'

Andrew nodded. So we walked through the schoolyard together. Children were running, laughing and shouting, just like any other day. We entered the school.

'Is Mrs. Pim around?' I asked one of the teachers.

'She's not in today.'

'Could I have a word with you then? In private?'

'Certainly. Just a minute.'

We waited outside the staffroom. On the wall was a map of the world. Andrew wanted to see where the plane had gone down. I showed him New York and Geneva. Then I showed him Newfoundland. 'Off the coast of Newfoundland,' it had said on the radio. I had to hunt. The exact location of Newfoundland was not even part of my geographical knowledge (nor was it in fact where the crash had occurred). Eventually, the teacher ushered me inside. 'My sister died last night in the Swissair plane crash,' I said. The words hung for a moment in the air, then evaporated.

'Are you certain?'

'Yes.' I was crying a little. I muttered something about this information being relevant to my son's behaviour, then I left.

I headed back through the wooded area around another

school. As I waited to cross the main road, the lollipop lady asked me a question. It barely registered with me.

'I'm sorry,' I said. 'I can't answer you now. I've had bad news from the States.'

'Your mother?' she asked casually.

'No. My sister. She died in the Swissair plane crash.'

The lollipop lady's face twisted. The change struck me. Sudden death by plane crash, apparently, was more significant than the loss of a mother.

I hurried home.

By now, it was a few minutes after 9 am. At the Dublin office of Delta Airlines, I got the answering machine. A woman identified herself as Liz, and asked me to leave a message. 'This is a difficult message,' I began, then explained that I believed my sister had been the Delta representative on the Swissair flight, and that I needed confirmation. Liz never rang back.

Swissair was next. This time someone answered. I gave my name, my address, and said what I wanted.

'That's nothing to do with me,' the woman said. 'You can call this number in Geneva.' She spewed out some digits and hung up. 'Unhelpful,' I scrawled on my sheet of paper. And as it happened, her number was wrong. Although I had clearly identified Patty as a crew member, she had given me the number for passengers. Geneva supplied a Zurich number, which I rang. Promptly, a woman confirmed in English that yes, Patricia Eberhart had been on Flight 111, and that there were no survivors.

'Thank you,' I said.

I did not cry, because I already knew. I had known for an hour. The death of my only sister, four years my senior, was already part of what I understood about the world. The conversation merely put the seal of authority on it.

The telephone was still in front of me. The most difficult call remained. 'What am I going to say to her?' I asked my husband, who had come home, having abandoned work.

Frank shook his head.

What can you say to a parent who has just lost a child? Especially when that parent is your eighty-year-old mother. My hand was on the phone, but I couldn't dial the numbers. I couldn't find the words just yet. Five minutes, I thought. In five minutes I'll do it.

Patty and Mother had been in Dublin the previous Christmas. My mother had insisted on giving me two packs of her Camel cigarettes, although I rarely smoked. But now there was a pit in my stomach, a big, dark hole. So I took a pack of cigarettes from the cupboard where they had been for nine months. I lit up and inhaled. Each drag of smoke flooded into the pit in my stomach, anaesthetising it, creating the illusion of betterment. I smoked the cigarette down to my fingertips, stubbed it out and dialled New York.

It was not yet 5 am in Mother's apartment. The phone began ringing. I pictured it on its triangular table, next to the chair in my mother's bedroom. I listened to it ring. It rang and rang. Deciding that I had misdialled, I hung up and tried again. I began counting the rings. Twenty. Thirty. It rang out.

There must be some mistake. Didn't I know that Mother was in her apartment? Had she not phoned me only an hour ago? Hadn't she left the message that was still on my machine, the message that Patty was dead? I dialled a third time. My mother was a little deaf, but she always heard the phone. Maybe she had fallen asleep in the living room, on the recliner chair, as she often did. If she was sound asleep in the living room, it was possible, just possible, that the phone in the bedroom wouldn't wake her.

But how could she have fallen asleep with the raw new knowledge of Patty's death?

I thought about my mother's heart condition. A few years earlier, she had gone in for a cataract operation, but the surgeon had refused to operate because her heart was fibrillating. I thought of the terrible shock of losing her daughter, and the eerie calm of the voice on my answering machine. Why was my mother not answering her phone? Had she collapsed in the apartment? Was she dead like my sister? Had her heart exploded in grief? And why had I waited for five minutes? Why didn't I know what to say to an eighty-year-old woman who had just lost her daughter? Making that phone call had seemed to be the nadir of despair, but this was even worse.

My mind lurched. What if Mother needed urgent attention and couldn't get to the door or phone? What if she was in sight of the phone ringing on the triangular table that I could picture so vividly? Who had keys to her apartment? I did, of course, but they weren't much use three thousand miles away. And Patty's keys were at the bottom of the ocean. There were no other key-holders.

I phoned again. The ringing was shrill, a controlled screaming in the dark apartment. Only the dead could sleep through such a noise. I was overwhelmed. When I had gone to sleep, I had had a sister and a mother. In a couple of hours, I had lost first one, and now, apparently, the other. And what could I do from across an ocean?

Was there anyone else? Any responsible, reliable person I could mobilise at this ungodly hour on my mother's behalf? I racked my brains. My mother had neither the inclination nor the temperament for friends. She did not go to church or to a club, and she found the company of women boring. She never chatted regularly on the phone to anyone except Patty and me.

Neighbours? Although she had lived in the same building for more than fifty years, her neighbours were acquaintances, not intimates. Relatives? She was out of touch with all her blood relatives, but on my father's side, there were possibilities, and a name came into my head. Considering the circumstances, Henry was not ideal, since he didn't live nearby. But at least he would be heading for work in Manhattan, and he knew us. Frank got the number from Directory Inquiries, and I dialled, picturing a house where I had never been, a house filled with young, sleeping children. Henry answered, his voice heavy with sleep. He stumbled over his words, scarcely believing what I was saying, then began to cry. What a terrible thing, I thought, to wake someone up with such awful news.

'It's Mom that I'm worried about. She's not answering her phone. I'm afraid that she's collapsed in her apartment, and nobody's got a key.'

Henry promised to do something.

I tried Mother again. Nothing but the ringing phone.

Obviously, I had to get to New York, and as soon as possible. But first someone special would be needed to help with Andrew while I was away, someone who could provide love in a difficult situation. I telephoned his godmother. Without hesitation, she agreed.

The phone rang. It was Delta Airlines, calling from America. 'We are very sorry about what's happened. We want you to know that we're here for you. Can we do anything?'

'Find my mother,' I said.

'Leave it with me,' the voice replied.

A few minutes later the phone rang again. It was – miraculously – my mother! The relief was overwhelming. 'Sorry,' she said, 'but I was downstairs in the lobby, talking to some people from Delta. They just wouldn't let me go. Finally I had to tell them to go home. Of course they wanted to come upstairs, but I wouldn't let them.'

Come upstairs? I nearly smiled. Of course she wouldn't let them. Mother loathed letting anyone into her apartment. 'I should have let you know,' my mother continued. For a moment, she seemed vaguely aware of the despair that I had felt when I couldn't get through. In the meantime, her voice sounded robust. She seemed to be her usual feisty self.

'Your sister and I were like an old married couple,' she said. The analogy startled me. It was Mother at her most articulate, welding language to insight. *An old married couple*. What better way to describe her intimate, turbulent relationship with Patty, their habitual bickerings? Gradually, the conversation drew to a close. All I could register was gratitude that they hadn't both been taken away during those few appalling hours.

Andrew's godmother arrived on the doorstep, brimming with support. I smoked a second cigarette while we talked, sucking each narcotising molecule of smoke all the way down to the terrible pit in the centre of my stomach.

The phone kept ringing. Most of the calls were from Delta Airlines: from their headquarters in Atlanta, Georgia; from their offices in New York; and later on, from Dublin. At least two individuals followed up on the first call. 'Your mother is with our personnel,' they both reassured me. The Delta people were insistently well-meaning, keen to let me know that I could reach them at all times. Each had three numbers: work, home and mobile, and occasionally a fourth number was on offer, that of an emergency headquarters at the Ramada Hotel, near Kennedy Airport. The people who called me all seemed to be keeping in touch with one another. I pictured busy phones ringing, ringing, ringing, crisscrossing America and the Atlantic Ocean, an industrious network of crisis management in full flow.

More than once, I asked about the recovery of Patty's

remains. I had such a clear vision of her body floating in the waves, that I assumed recovery and identification would be prompt and straightforward.

⚜

On my computer there was an email from Richard. He had sent it as soon as he had gotten to work: 'CNN's web site reports that just one Delta flight attendant was on board the flight.' I knew what Richard was thinking: just one. With thousands of flight attendants, how could that 'just one' be Patty? I emailed him briefly: 'The news is bad. Try to keep happy. Love, Mom.'

He was on the phone within a minute. 'Is it certain? Is it definite?'

'I'm afraid so. Be cheerful. What's the point in being anything else?'

⚜

The pit in my stomach deepened. I smoked another cigarette, my third, drawing in each puff as far as it would go. I went for a walk around the block. It was mild and bright, a lovely day. A charm glittered at my feet. I picked it up. It was an inexpensive little thing: two tiny white doves with spread wings, perched on a double silver band. What did it mean? Two doves, fluttering on linked rings? Why had I found such a thing on such a day?

Back at my desk, I made a list of what needed to be done. It was a long list, and the phone kept ringing.

The day vanished. Frank and I went to collect Andrew, walking hand-in-hand towards the school gate. A recently separated mother stared sardonically. Lovebirds! I could see her thinking. Andrew, however, was delighted by this unprecedented event: both parents collecting him. He chattered without reference to his aunt, but his good humour vanished when he walked into the house and smelled the cigarette smoke. At once, he hunted out the ashtray, discovered the three butts, and became outraged.

I tried to defend the indefensible. I was upset, I explained. The few cigarettes were doing me good. Under the circumstances, he should be a little more tolerant. Foolishly, I lit up. Andrew was unimpressed. He stamped and raged. As far as he was concerned, an individual cigarette was as bad as the habit. Right away, I stubbed out the cigarette, but Andrew would not speak to me. Slowly, I wrapped my mind around the fact that his distress was about more than cigarettes. It was also about the awful thing that had happened. Andrew's nine-year-old world had become unpredictable, and he was frightened.

❦

'No, I can't travel tomorrow,' I said to Delta's senior representative in Dublin, who had pencilled me in on the next day's flight. My list was long, and still growing.

'Saturday, then,' he said, suggesting a return flight five days later. His voice had a solid American drawl. He sounded large and clean-cut. I could picture his short hair, his dark suit and white shirt, his clean-shaven face.

A return flight five days later?

My mother was eighty years of age. She had heart trouble. Her breathing was roughened by six-and-a-half decades of cigarettes, and she had had a touch of skin cancer. Her beloved first child, with whom she found completion as 'an old married couple', had just perished violently and unexpectedly. Her husband had died eighteen months earlier, after years of dementia. And I was to pop in to offer a few days' consolation?

'My mother is going to need a lot of support,' I said politely. 'I'll return in a month.'

'Fine,' the man from Delta said. 'And would you like to bring anyone with you?'

Unknownst to him, the question was an awkward one. Mother's objection to people in her apartment included my husband. He had slept there only once, twenty-two years before. Immediately afterwards, my mother had thrown out the spare bed. 'You married him,' she had said. 'I didn't.'

I realised that Delta was offering to put Frank into a hotel. Grateful, I suggested dates for his travel. My mother had already insisted that we weren't to travel on the same plane.

                          ✦

I made dinner for the boys. My mind was filled with the difficult task of telling Frank's mother what had happened. Sybil, a widow, was hospitalised with depression, and the news of Patty's death was likely to distress her. As Frank and I drove to the hospital, I had an idea. We could try putting it to her in a way that might pull her out of herself.

Sybil was in bed. 'We're going to need your help,' I said. 'You're going to have to be strong for us.' We smiled, holding her hands, and asked if she had watched the news on television. She had. I said it out then, that Patty had died in the plane crash. Sybil struggled to understand, to relate the television images to my sister, whom she liked. As the realisation sank in, my mother-in-law began to weep, a pathetic, weak sobbing.

We stayed a while longer, talking cheerfully. Before leaving, we asked the nurse to keep a watchful eye on her.

On the way home, we stopped for a meal. Dinner in the house would have been punctuated by the phone ringing. An hour together, beyond reach, was an oasis.

                          ✦

Andrew was in bed. I gave him a kiss and a squeeze, snuck downstairs, and surreptitiously lit another cigarette, my fifth. I needed it. I was about to watch the news. I inhaled like a vacuum, sucking in the smoke as far as it would go. They were talking about smoke on the television too. Minutes before the crash (which had occurred, in fact, off the coast of Nova Scotia), the captain of the MD-11 had reported smoke in the cockpit. Those who had heard the plane were interviewed. One described the scream of ailing engines. Another, the sound of the impact, a shattering blast, although the plane had come down five miles offshore. Then a short clip flickered before my eyes: men in orange jackets unloading a body bag onto a wharf.

Was it Patty? I wondered.

Exhausted, I fell into bed and slept at once. Soon I was awake again with cramps. I got up and moved my bowels. My stomach still hurt. I dozed. A few minutes later, I was up again, amazed by the copiousness of what I produced. My gut was sour; there was an unpleasant chemical burn. The pain prevented sleep. I tossed and turned. All I could think about was the image of a huge plane settling into the sea. I saw the cold, black water seeping in; I felt the silence and the dark.

In the toilet, I expelled yet more foul sludge, quantities of it. Yet nothing relieved the sour, chemical burn.

Back in bed, I switched on the radio, using my earphones so as not to disturb Frank. I lay there in pain, listening to the soothing tones of BBC's World Service. The news came on. The Canadians, I heard, had still not given up hope of survivors. I thought of Patty: of her dogged persistence; her wilfulness; the unbending strength of her character. She gave nothing up easily. She would have hung on to life until the very nanosecond that it was wrenched away. But she was dead. I had no illusions, no hopes. They were all dead, every man, woman and child who had boarded the plane in New York.

'No survivors,' the woman in Zurich had said, after she had

confirmed, six hours after the crash, that Patty was on board.

Where, exactly, were the Canadians looking for survivors? Did they imagine that a herd of dolphins had rallied to the scene, towing passengers to shore? Or did they conceive of the submerged aircraft as a giant air pocket, which allowed people, warmly wrapped in Swissair blankets, to breathe beneath the waves? Yet what else could they do, but put a gloss on it, look for hope in the face of hopelessness? 'Try to keep cheerful,' I had said to Richard. 'We're going to need your help,' I had told my mother-in-law.

I went to the toilet again. When I got back into bed, I was shivering uncontrollably. I was furious with myself. How dare I shiver with cold when I was alive, in my own bed, in my own house when, only the night before, my sister had died in the cold, black sea? Now *that* would have been a slimy, terrifying cold, the kind that would squeeze the heart and burst the lungs. So what right had my pampered body to complain?

I got up again. It was hard to believe that the length of my gut could hold so much. What's more, the chemical burn was getting worse, spreading upwards through my oesophagus into my mouth. Suddenly, I figured it out. The cigarettes were to blame, the five cigarettes that I had smoked to anaesthetise myself. Their nicotine and other chemicals had provoked my healthy body to defend itself, and expel the poisons.

Andrew's deep sense of justice would be satisfied. He would say that it served me right.

FRIDAY, 4 SEPTEMBER

Knowing that I had been awake all night, Frank took Andrew to school. I fell asleep at last, and was out for the count when the phone rang. It was the rector who taught religious education at Andrew's school. He was full of sympathetic questions, but answering proved difficult. My words were squeezed out in a thick, slurred voice. I told him that Patty had been working on the flight as a senior attendant, and that twenty minutes had elapsed between the first report of smoke and the crash. Already, I was focused on the notion that activity had absorbed her. How much better, I thought, to be busy helping others, than to be strapped in a seat waiting to die.

I asked the rector what time it was. I thanked him for waking me up; my list was waiting.

I was still cold. I switched on the heating, climbed back under the covers with the telephone, and began to extricate myself from my teaching commitments. At the same time, spoonful by medicinal spoonful, I munched through a small bowl of cereal, my sour stomach protesting. With each phone call, my grip on reality weakened. Everything that I said about Patty's death sounded more and more bizarre. How could this appalling thing have happened? Why was I saying that it had? I grew colder and colder. I had only to put the phone in its cradle, and it would ring again: friends offering consolation and help; or people from Delta, attempting to do the same.

The doorbell rang. I put down the phone, opened a window and saw my friend Liz downstairs. She was clutching a potted plant. I let her in, then went back to the phone. By the time I joined her, she was making tea.

Andrew telephoned from school. Had I remembered that today was a half-day?

No, I had not. I pulled on some clothes, and Liz drove me to the school. I noticed a second plant on the back seat of her car.

'Somebody else?' I asked.

'The wife of a friend. In a car crash.'

I shivered. Death was all around me. How was it that I had never noticed before? Flocks of living birds sweep the sky, but I had seen dead birds only infrequently, and never in pairs, let alone multitudes. I had lived for nearly half a century, yet never witnessed the moment of death of a fellow human being. But now I felt death everywhere, even in the air I was breathing.

❧

The afternoon disappeared in necessary arrangements. I was trying desperately to complete them, when a friend of Patty's telephoned from America, incoherent with grief. I tried to console her, but hadn't much energy. Her sobbing was infectious, but I couldn't afford to dissolve with her. I had to remain in control. The business of coping was taking everything I had. The call dragged on, then thankfully, it was over. I made dinner, did laundry, put Andrew to bed. Some dear friends called to the door. We sat down for a few minutes, then I sent them home. I still had to pack. I was going to New York in the morning.

It was late when the senior representative of Delta in Dublin telephoned, his fourth, or perhaps fifth, call of the day. Not for the first time, he asked if he might accompany me on my flight. I had already declined. This time, the way that he asked, it would have been churlish to refuse.

SATURDAY, 5 SEPTEMBER

My bag was ready, with the usual, panic items tossed in on top of my orderly packing from the night before. My change purse was full of heavy Irish coins. I sorted them into two heaps, handing each to a son, according to their needs.

The man from Delta was on the phone. What time would I be arriving at the airport? He would be waiting at the entrance to the terminal.

On the way, we stopped in the centre of Dublin. I needed tissues (just in case) and something for my stomach, which was still in an uproar. An ordinary Saturday morning would have found me in exactly the same place, walking down Nassau Street bound for the National Library. Under the circumstances, the normality of the prospect, seemed very peculiar.

In the car again, we headed north, across the glittering Liffey. Andrew was growing sadder by the minute. 'I don't want you to go,' he wailed.

'I don't want to either. But I have to.' All the way to the airport, we composed a symphony of variations around this inescapable motif.

At the set-down area, we hugged and squeezed one another. Then husband and son were gone, leaving me with the image of my child's tears. Hoisting my bag onto my shoulder, I entered the terminal. At once, I picked out my minder. Far from the perfect specimen of American manhood that I had anticipated, he was neither tall nor handsome. In fact, he was paunchy and balding. He was not alone. Another man and two women were scanning the crowd with him. I had supplied a description of myself and the clothes I would be wearing, but they still did not see me. I passed through security, circled back and stood at an angle, watching them.

I was glad to be invisible. I embrace anonymity; to be known limits the possibilities. For a few more seconds, I cherished my privacy. Then I finally said, 'Is that you, John?'

He whirled around to face me, his mouth a startled O of surprise. I was introduced, shook several hands, was told how sorry everyone was. And then, because my sister had died a hideous death in the course of her daily work, I was treated as a Very Important Person. My passport and suitcase were whisked away; somebody checked me in. I was taken to a courtesy lounge, where tea was brought in a silver teapot. I opened the Irish newspapers to find reports about the air crash. The force of the impact, I read, was such that most of the bodies had disintegrated; it was pieces, rather than bodies, that were being collected from the sea.

I sipped my tea. Pieces, not bodies. My mind absorbed this. If there were pieces and not bodies, then my images – of the aircraft sliding into the sea; of cold, black waters encroaching; of stunned, slow drowning – must be false. If there were pieces, not bodies, then the MD-11 must have hit the water with unimaginable force. The victims could not have died slowly, not if they had ended up as pieces.

The man from Delta came back with my passport, ticket and boarding pass. We would be travelling first-class on Aer Lingus, which, like Swissair, enjoyed a code-sharing agreement with Delta. I knew perfectly well why he had insisted on joining me: no airline wants a bawling woman disrupting the calm of an aircraft, or a woman who might remark to a fellow passenger that her sister's body parts were floating in the Atlantic, following an air crash. As it happened, I was in control of myself, more or less.

I stared out of the window during the take-off, watched the familiar sights recede beneath me. A fug of cloud embraced us, then we burst through it into sunlight. The seats were roomy and comfortable. Papers were brought, a lavish array of English, Irish and American dailies. I took two English papers, *The Times* and *The Independent*. I immediately turned to the articles about the crash.

Pan-pan-pan. The monosyllables jumped off the page, an international distress signal, derived from the French word *panne,* meaning mechanical breakdown. They were the first sign from the pilots of Swissair 111 that all was not well. To air traffic controllers, the words signal a technical problem, and the desire to land as quickly as possible. They do not indicate a full-blown emergency. The word for that is *mayday*. *Pan-pan-pan*. The words hooked into my imagination. They rang like a refrain from a Christmas carol, like a nursery rhyme, like a child's laughter, like anything except what they were: the harbinger of death.

I moved on from the crash articles. I pored over the papers, reading everything, even the sports, occupying my mind with words, with ideas, with anything that would crowd out the one thing that was preoccupying me. But as I read, the three syllables punctuated each paragraph relentlessly: *pan-pan-pan*. A playful, deadly patter. Linen cloths were spread in front of me. Crystal glasses of orange juice and water were replenished. Canapés were brought, then an attractively presented meal. I was asked what kind of wine I wanted.

I demurred. What kind of *wine* did I want? My sister had just died, I was flying to New York to be with my bereaved mother. The thought of applying my mind to selecting wine seemed inappropriate.

Two new words popped into my head: *blood money*. *Blood money* was the expression my mother had used, thirty or forty years before, when her father, our grandfather, had given Patty and me money for our birthdays. He was elderly, unhealthy and an alcoholic. He earned his crust as a nightwatchman. We were to take his money with respect, my mother was saying. We were to understand that he had earned it with his blood. All too clearly, I understood that it was *blood money* that had installed me in a spacious leather seat in the premier section of Aer Lingus. It was my sister's blood that had placed the crystal glass in my hand and the fillet steak on my plate. The orange juice slithered down my throat. The food proved more problematic. My insides were knotted, my bowels on overdrive. But if I did not eat, how could I function? So I cut and chewed, allowing myself to stop when half the food was gone.

I got up to use the toilet. The cockpit door was open. Radiant light streamed through the spread of the windows, silhouetting the backs of the two pilots. The compact space was bright with instrumentation. To my unaccustomed eye, the cockpit appeared as a world of its own, encapsulated, shining with the unfamiliar, a world unknown to the ordinary traveller. Now I was conscious of it being part of Patty's world, a space with which she had been intimate. My eyes filled with quiet tears. What little interest I had taken in Patty's work, how little I had even thought about it. Now that it was too late, I yearned for a better sense of her life, a chance, however superficial, to relate to her perspective.

When I returned to my seat, I asked John if I might visit the cockpit. A few minutes later, I was sitting behind pilot and co-pilot. The pilots, like the rest of the crew, knew why I was travelling and, as I sat with them, I became aware that their minds were as powerfully exercised by the fate of Swissair 111 as mine was, although for different reasons. So I asked about the smoke in the cockpit. It might have come from anywhere, I was told. Smoke can float along the ducts from all over the aircraft. 'But in this case,' the pilot said, 'the smoke probably came out of the panel behind you.'

I stared at the bland, smooth surface.

'That's the electronics bay. It's packed with instrumentation of all kinds.'

'Would the pilots have lost control because they were incapacitated by the smoke?'

'No, they'd have had oxygen masks. The problems start when the smoke gets too thick to read the instrumentation. If the pilots can't read altitude and airspeed, they can't fly the plane. So they'll try to get low enough to open the window, which will clear the smoke.' The pilot tapped on the window beside him. The idea that it could be opened startled me. 'Of course,' he added, 'with an open window, you've a problem with noise.'

I asked what my sister might have been doing during the last minutes.

'Very busy,' the pilot assured me. 'She'd have been clearing the aisles and giving emergency instructions.' In my head, I pictured Patty's slight figure struggling with a heavy trolley, fighting the judder of the aircraft. I said something, then, about how I had imagined that water would cushion an out-of-control aircraft, that the sea was better for a crash landing.

The pilot shook his head. 'Naw,' he drawled. 'It's like hitting concrete.'

Like concrete. The word was solid and unanswerable. I returned to the cabin, overwhelmed, but reassured by the idea of instant oblivion.

⚜

Back in my seat, I spent an hour or so asking my travelling companion about himself. Obligingly, he rattled on. As with reading the newspapers, it was a way to keep the horrors at bay.

About twenty minutes out of New York, John was told that he was welcome in the cockpit for landing. He turned to me. 'Would you like to take my place?'

In seconds, I was harnessed in behind the pilots. A few minutes later, John materialised as well, scrambling over me into the corner.

There is an intimacy in a cockpit. Leaning forward slightly, I might have propped my chin on the co-pilot's shoulder. The vista was astonishing. As the aircraft banked, I had a sense of the entire world revealing itself: the sky; the sea; boats and their wakes; the intricate topography of the land and man's constructions. The pilots were no longer their easy-going, mid-Atlantic selves. I could sense their concentration, their focus on getting things right. Instructions hammered in from air traffic control. The pilot repeated each as he implemented it, under the scrutiny of the co-pilot. Information was supplied about other aircraft in the vicinity. The pilot zoomed up and down in his chair, visually checking their positions in relation to our own. Then the massive, blank expanse of the runway stretched out in front of us. In the distance, Manhattan rose out of the haze. We were pointed directly at the Empire State Building. The details of our altitude punctuated the air: 50 feet, 40 feet, 30 feet, 20 feet, 10 feet. We touched ground. The atmosphere in the tiny nose of the jet was intense. Flung forward, I was simultaneously pulled back by reverse thrust. I was glad of the harness, and John's arm, which he had stretched across me.

Frightened and thrilled, my heart was beating fast. Never had I imagined landing at JFK in the cockpit of an airbus. *Blood money*, I thought. Again, it was Patty's horrible death that had brought about such a privilege.

I remained in the cockpit as we taxied towards the terminal. My emotions were in conflict: high from excitement, but also distraught, mirroring the push-pull of the landing I had just experienced. The door of the aircraft swung open. Five people were bunched outside it, half-obstructing the way. 'That's her,' I heard someone say. My heart sank. This reception committee had an advantage over the one in Dublin. These people had known Patty, and they recognised her features in me.

I didn't feel much like talking to anyone, let alone strangers. And yet, these were decent people, intent upon doing the right thing. One of them, as I found out later, had flown up from Atlanta to be there. I shook everyone's hand. The name of one woman was known to me. A man's face looked familiar. Surely Patty had introduced him on some past occasion in the airport. But I would have known neither had I passed them in the street. My travelling companion, a comparatively old friend, had disappeared. I was under siege. What could they do for me? Was I hungry? Would I like a drink? Did I want to stop anywhere on the way in from the airport? I looked at their kind faces, focused on the impossible task of making the unbearable better. What could I say that would fulfil their need to serve?

'I'd like to call my mother to tell her I've landed,' I said. 'Could I please use one of your mobiles?' Hands reached to pockets. The five exchanged looks of consternation. It had seemed safe to assume they all had mobile phones. And yet here they were, addled, because not one had a phone with them. I was ushered away to an office where I made the call.

In the meantime, someone had fetched my bag and cleared it through customs. One of the women was going to drive me into the city. When we got to the car park on the roof of the Delta terminal, she vanished.

'Where did she go?' I asked.

'Over there. She's in a hurry today.'

The woman had run for her van to save me the trouble of walking across the car park.

It was mid-afternoon when we pulled up in front of my mother's apartment building in Manhattan. Being a Very Important Person had shaved ninety minutes off the journey from the airport. Invariably, powerful emotions hit me as I catch sight of the building. Its yellow brickwork, its awning, and the familiar doormen, merge reassuringly with memory. I was probably conceived in the building. I learned to walk there, went to school from there, kissed the occasional boyfriend up on the rooftop. After thirty years in Ireland, I still called that apartment home, exactly as I called my house in Dublin home. But now, something was different. This building, this physical setting of our shared past, was no longer a place where I would meet Patty. My sister's violent death had propelled me into a world where familiar things had changed forever.

I said goodbye to the woman from Delta, thanking her for the seamless journey into the city. 'I want you to know that we're here for you,' she said.

I nodded. 'I just want to be alone with my mother for a while.' I lugged my bag down the steps. At the reception desk, two enormous bouquets were waiting to be taken upstairs. I slung the bag over my shoulder and scooped up the vases, scarcely able to see through the forest of blooms.

Upstairs the apartment door swung open. My mother gasped at the flowers. 'You'll have to do something about those right away,' she said. Mother hovered as I fiddled with them. The smoke from her cigarette drifted towards the kitchen ceiling in a blue-grey spiral.

'I was sitting in the living room when she died,' she said. 'I was cutting out articles for her to read. I fell asleep. Later I went into the bedroom. It was 2 am. I switched on the radio. "Flight 111," I heard. I realised at once that this was it. This was the Big One. I didn't cry and I didn't scream. This was the Big Enchilada.' My mother's voice was collected. I followed her into the living room, where she settled into the recliner chair surrounded by newspapers. I cleared off the chair next to the fireplace, so that I could sit too. On the floor beside me was the shopping bag which contained my father's cremains. She had left it there, next to the fireplace, since his death.

'"We'll do Connecticut on Saturday,"' Mother said, referring to the house that she owned on the Connecticut shoreline. 'That was the last thing that my beautiful Patricia said to me. And now we'll never do Connecticut again.' Dreamily, Mother lit another cigarette. 'I do not want her remains,' she said. 'I told Delta to leave them at the bottom of the sea.'

'That's not the way it works, Mom,' I said. 'They'll bring up whatever they find.'

'But what am I supposed to do with them?' Mother said, waving at the shopping bag that contained my father. 'Stack them up?'

After a while, I got up and walked around the untidy apartment. The mess in the kitchen was the worst. The floor was scarcely visible beneath the clutter of shoes, bags and reading material. The table, long since abandoned as a place to eat, was covered with newspaper clippings, correspondence and financial documents. On the chest of drawers was a jumble of cake boxes, corn chips, condiments. A cellophane bag lurked dangerously close to the ancient toaster oven. Near the window, bottles of bleach and ammonia mingled with soft drinks. The counter was obscured by cans and boxes, and hundreds of the tiny packets of salt and pepper that get thrown in with takeaways. Bulging plastic bags dangled from cupboard doors.

Cautiously I opened the fridge, which was packed with brown paper bags and styrofoam containers. I poked about, unearthing items that I had seen months earlier. Once I had thrown out a piece of raw liver discovered in my mother's fridge, its plastic bag date-stamped three years earlier.

In the bedroom, which Patty and I had shared as children, bags and boxes were stacked along a wall. The single bed was heaped with newspaper clippings. 'I'll have to move this stuff,' I called, 'if I'm to sleep tonight.'

My mother came along. 'But they're for you,' she said. 'I want you to read everything on that bed.' She picked up a newspaper clipping. 'Ah,' she said, 'this is wonderful. *Wonderful.*' I looked over her shoulder. The clipping was a book review. The book was called *At the Water's Edge: Macroevolution and the Transformation of Life.* 'That's some article,' Mother said. 'If you read that article, you'll understand everything. You'll understand that your sister was a successful metazoan.'

She handed it to me. 'What then is macroevolution?' I read. 'Is it anything more than microevolution left running for a thousand or a million generations?' The words swam in front of me.

'I want you to get ten copies of that article made,' my mother said. 'Your husband should read it. Everybody should read it. Yes, Patricia was a successful metazoan.'

In the bathroom, plastic bags hung from the doorknob, making it impossible to close the door. The shower was occupied by clothes on a rack, cans of paint, buckets and brushes. Around my ankles were boxes and bottles, and the ubiquitous newspapers and books.

There wasn't an inch to hang my jacket.

⚜

As always, we went out for dinner. My mother and I were climbing the steps to the street when someone called me. I spun around. There was no one there.

Yet I had heard my name being called, as clear as a bell.

Mother had something to copy, so we stopped in a newsagent's with a self-service photocopier. As we paid, the rack of tabloids was impossible to avoid. All led with searing headlines about the air crash. I picked up *The Post*. My mother frowned. 'I lost my other daughter on that,' she told the man behind the counter.

'What?' he said. 'I'm so sorry. So sorry.'

We ate in a deserted diner on First Avenue, the kind of inexpensive place that my mother favours. 'You shouldn't drink Diet Coke,' she said. 'It's bad for you.' I ate with effort, my battered gut complaining. By the time we finished, my mother had made friends with the waiter by successfully guessing what part of India he came from. He was both astonished and pleased.

The fact is that my mother knows something about everything. She has read it all in the newspapers.

Back in the apartment building, I trawled the refuse rooms, wandering from floor to floor until I found the day's *New York Times*, already disposed of by an efficient reader. Although a hoarder of newsprint, my mother disapproves of buying newspapers. Instead she collects quantities of out-of-date newspapers from the refuse room nearest to her apartment. Of course I could have bought the day's paper myself, but then Mother would have complained about my extravagance.

Studying the articles about the crash, I learned that the plane's nose had dropped during its final moments. The thought dismayed me. How long did it last, that final, terrifying plummet? Thirty seconds? A minute? I saw Patty, her feet flying away from the floor of the cabin, her head slamming into its roof. I looked at the photographs of some of the people who had died with her, but they scarcely registered. For the moment, I could think only of Patty. My mind was filled with questions. All that I read seemed superficial. I wanted to know everything there was to know about Patty's death.

'I'd like to go to Nova Scotia,' I said to my mother.

'What would you want to do that for?' she said, making a face. 'I forbid it. I absolutely forbid it.'

Unable to use the shower because of the junk inside it, I headed for the bath. As soon as I was naked and soaking, the bathroom door creaked open.

'Hey!' I complained.

'I'm not looking,' Mother said. 'Just don't forget to make ten copies of that article I left on your bed. Remember, your sister was a successful metazoan.'

What in God's name was a metazoan?

⁕

It was 10 pm in Manhattan, 3 am in Dublin. I was clean, but tired. After removing the newspaper clippings, I crawled into bed. I slept the sleep of the exhausted. At two o'clock, sharp cramps woke me. I fled to the toilet. Afterwards, I looked in astonishment. What I saw was plentiful, and I had hardly eaten anything. I slept fitfully for another two hours, before cramps got me up again. Clearly my insides weren't interested in sleep. I was wide awake, my brain on overdrive. I took out my diary and began to write.

SUNDAY, 6 SEPTEMBER

7 am. Beneath my window, the city was stretching into wakefulness. Already the bedroom was warm. I was glad of the heat. My hands and feet were cold. In spite of the warmth, I was ice-cold within my skin. The radiator by the window was covered with detritus. A pamphlet caught my eye: it was entitled, 'Welcome to the Swissair Cockpit'. I opened it and read:

> Dear Passenger,
> Thank you for coming to visit us in the cockpit. We are pleased to see you're so interested in flying. We would like to answer your questions, since flying is not just our job, but our hobby, too. Unfortunately, especially on short-haul flights, we are often busy navigating, communicating, monitoring, or just being prepared for whatever might happen. So we've compiled this brochure, to answer some of the questions which we are most frequently asked.

My eye scanned the questions and answers, zeroing in on Number 10: How safe is flying? The lengthy answer described extensive back-up systems, highly qualified maintenance people and dedicated cockpit management. 'Flying is the safest way to travel,' it concluded.

Pan-pan-pan.

✤

Cautiously I ate: cereal and Japanese tea, brought in my bag. I ate standing, because I couldn't sit, on the cluttered kitchen chairs. Then I tiptoed into my mother's bedroom. She was asleep on top of the bed. With sleeping, as with everything else, my mother has her own system. She sleeps on – as opposed to inside – the bed to avoid making it. She sleeps late because she stays up late. She starts her sleep in the recliner chair in the living room, then finishes it in the bedroom. Even as children, we never saw her until the afternoon, because she was asleep when we left for school.

Since Mother was asleep, I could clean the fridge. Swiftly, I removed a selection of half-eaten rolls, baked potatoes like rocks, Chinese leftovers in yellowing cartons. Next I tackled the mess on a chair, so I would have a place to sit. Then I went out, taking the refuse with me.

In the downstairs lobby was a doorman I didn't recognise. 'I'm Mrs. Eberhart's daughter,' I said to him, then my eyes burned. *Mrs. Eberhart's daughter.* In fact, I was now Mrs. Eberhart's *only* daughter.

Outside, I began to walk as fast as I could, but striding through those beautiful, Sunday morning streets brought no pleasure. All I could think about was Patty. With each step the realisation hammered home. Never again would we walk anywhere together. I stopped in front of an exquisite window of needlepoint pillows. Each bore a message. *Mrs. Wonderful*, I read. Patty would have loved it. She would have wanted to buy it for Mother. Another read, *Anyone Who Says Money Doesn't Buy*

Happiness Doesn't Know Where to Shop. Why, that statement *was* Patty. I wanted to break the window and grab the pillow. But to what end? Patty was dead. Who else would know exactly why such a thing could bring us together for a fleeting moment of understanding?

I marched on purposelessly. I ended up in a coffee shop, listening to Puccini, letting the hot coffee scald my mouth and throat, as if physical pain could make things better.

Back in the apartment, I decided to go to the Episcopal church where Patty and I had been baptised and confirmed, and where we had both taught Sunday School. However Mother, now awake, told me to go to the Brick Church instead, which Patty had attended sporadically as an adult. So I walked uptown. Arriving early, I was ushered out into the Garden Room. There is something exquisite about gardens in Manhattan, and none more so than the Brick Church's, all light, greenery and birdsong, amidst the soaring buildings. I sat quietly, breathing in the green-washed air, but the beauty could not dispel the images in my head: the sea boiling as it swallowed an aircraft; the stench of fuel; wreckage and remains bobbing to the surface.

Soon someone I knew came hurrying across the garden. It was Edgar, a lovely man to whom Patty had introduced me a few years before. We went into the church. As I settled with Edgar in a pew, a tearful woman in a yellow dress joined us. She introduced herself as Elizabeth, a friend of Patty's. And so we sat, three together, as the pastor directed some of her words and prayers around Patty, and the others who had perished with her. The formal service, with its particular references, gave shape to the chaos in my mind. It provided a spiritual acknowledgment of Patty's death, and I was glad of it. For I knew that my mother would neither plan nor sanction either a funeral or a memorial service for Patty. Never a believer, my mother has grown more and more outspoken in her dismissal of God and an afterlife.

Since my father's death, Mother has told me what to do after hers. I am to call Cremations Direct. I am to mix her cremains with Daddy's. I am to scatter them both in my garden in Ireland. There is not to be any funeral. 'If I had a funeral, people that I don't like would come,' she has said. Now, as I sat in the Brick Church, it struck me that my sister had never been included in these discussions. Why not? I wondered. And why Ireland? Was it some darkling instinct that whispered that Patty would not be there?

After the service we returned to the garden, where a number of kind people swayed in and out of my orbit. But when the pastor proposed making a call to my mother, I deflected her circumspectly.

Elizabeth told a story that was vintage Patty. For a recent party, my sister had handed her a little parcel. 'For distribution as party favours,' Patty had said. Inside were sixty fortunes, retrieved from fortune cookies, each one pleated into a miniature fan. Who else but Patty? Who else would collect sixty scraps of paper, then painstakingly pleat them for recycling?

'My feet are killing me,' I said to Mother, as I came through the door and kicked off my high heels.

'Those are winter shoes,' she said. 'You shouldn't wear shoes like that in heat like this.'

There were more flowers, armloads of them. As I attended to them, I gave her a detailed account of the church service and everything that had been said about Patty. My mother laughed at the sixty fortunes, then launched into a Patty story of her own. 'She was beginning to think about me dying. She said, "I don't want you to die, but just don't give me any trouble." She was most insistent. "You're not going to give me any trouble, are you, Mother?"'

And as my mother repeated those wry, charged words, I could hear Patty's voice.

I lay down. I wanted to be alone. I needed silence, nothingness. An image came into my head: Patty running down the aisle of the aircraft. She was upping her pace to fulfil her duties. Then another picture came: a white rose on the table in the foyer of Patty's building. Mother had said that a rose had been left there, with a card that read 'You will be missed.' It was signed M.

I got up from the bed. 'Marcio,' I said to my mother. 'Marcio was the M who left the rose.'

'Could be.'

Marcio was the driver who took Patty to the airport. I called him. 'I took her to the airport that day,' he said. He sounded miserable, as if he wished that he had crashed his van, and they had never made it. He sounded like he never wanted to drive anyone to the airport again, just in case... I thanked him for the flower. He said he hadn't been the one to leave it. I thought he was going to cry.

⚜

For dinner, my mother chose Kentucky Fried Chicken. Under bald lights, at a table studded with the crumbs of those who had dined before us, we ate with plastic forks out of boxes. Afterwards, we went for a walk through the dark, warm city streets. Nothing seemed real. We might have been wandering through a dream.

Back in the building, I wandered from floor to floor foraging for the day's *The New York Times*. When I returned to the apartment, Mother asked: 'Did you take the lift?'

'I did.'

'Well, don't,' she said. 'The doormen watch on their security cameras, and I don't want them to think you take newspapers out of the refuse rooms.'

I turned to the Swissair story. I read that the aircraft had lost contact with air traffic control six minutes before disappearing from the radar screens. Six minutes! I felt those six minutes stretch, each second an aeon. When they lost contact with air traffic control, I thought, the pilots must have despaired.

Information swirled in my head. I struggled to piece it together. The crisis was twenty minutes unravelling. Just when did Patty smell the smoke? When did she feel afraid? For how long – if at all – did she believe that the plane might land safely? And what did she do during those final six minutes? Six long minutes in a black, smoky cabin, in an airliner that was out of control? A limited transcript of the dialogue between the pilots and air traffic control was printed in the paper. I attempted to make sense of it. I lay on my back, reading and thinking, trying to get into the airplane, to be there with Patty, to share her fear and suffering, to make it less. When I went to bed, the images stayed with me, going round and round on a loop.

All night, I lurched in and out of bed, my bowels disgorging their foul contents, again and again.

MONDAY, 7 SEPTEMBER

7 am, Labour Day, a holiday in America. The fifth morning since Patty's death. I had started to think about the 228 other souls on board. One flight attendant had been on her first trip with Swissair. How could anyone die on a first trip? Patty, flying for thirty years, had survived several thousand flights. Later I would read that the chance of dying in a commercial airliner leaving an American airport is one in six million, far less than the risk of dying in the vehicle that takes you to the airport. But as with all

risk statistics: what do they matter when the victim is you?

More flowers arrived. Mother admired them, bloom by bloom. 'When I look at a flower,' she said, 'I want to become a botanist. But life isn't long enough for most things.'

I had brought photographs with me, pictures that my mother had never seen. Among them were two close-ups I had taken of Patty. Mother was thrilled; Patty's death had made them precious. A reporter from *People* had been asking for a picture, and Mother decided that she should have the better of the two. At the same time, she was reluctant to let it out of her possession. An arrangement was made to have the chosen photograph collected from the lobby downstairs. Then Mother dug in her heels. 'You can't trust the doormen,' she told me. 'You'll have to go downstairs yourself, and wait for the courier.'

When my mother decides that something must be done in a particular way, there is no dissuading her. So, while protesting that an exact time had not been arranged, I headed downstairs with the photograph, my diary and my upset stomach. By 10:15, the heat in the lobby was intense; the forecast was for 94 degrees. After two hours, I got fed up. The hot velvet of the chair was scratching me; I was soaked with perspiration. I retreated upstairs, telling my mother that I had better things to do than pass the entire day in the lobby.

Hours later, when the courier arrived, I went downstairs and gave him the precious picture.

My mother, who has theories about everything, wove a narrative around the two photographs. 'When you took the first one,' she said, 'you were just taking a picture. But when your sister saw that you wanted a second, she relaxed. She realised you were sincere, that you really wanted a photograph. And so, you got the real Patricia. My lovely Patricia.'

In the evening, an encounter by telephone. An officer from the Royal Canadian Mounted Police, named Dan, would be asking about Patty with a view to identifying her remains. An apology was tendered in advance: some of the questions would be personal, but all, I was assured, were necessary. Since there was only one phone, the plan was that I repeat each question verbatim, so Mother might contribute if she wished. The first questions were about me. Who was I? What was my relationship to the deceased? My date of birth? Just how would my date of birth assist in identifying my sister, I wondered, but I answered anyway. Dan chuckled. 'We're the same age,' he said cosily, as if this somehow made our bizarre conversation more acceptable.

Eventually he moved on to my sister. How old was she? How tall was she? What did she weigh? What colour were her eyes? Were they both the same colour? Were her ears pierced? Was any other part of her body pierced?

Not until this crash, I was tempted to say.

Did she have any tattoos? Did she wear an ankle chain?

I answered scrupulously, all the while thinking how these questions would have infuriated Patty. Like Mother, she was hostile to anyone knowing anything about her. And she would have been outraged by the mention of her name in conjunction with body-piercing, tattoos and ankle chains. Who was this Dan, I thought, and why hadn't he given me his surname? Were his eyes the same colour? And just where was his body pierced?

Dan asked about moles.

Moles? Mother and I discussed moles. Patty definitely had one on her shoulder, but was it the right or the left? I remembered seeing the mole, which was large and dark, and being shocked by it. 'You should have that checked out,' I had said.

Patty had stared at me blackly. 'Should I? Do you think it

47

would be enough? Enough to do for me, if I left it alone?'

I told Dan that we thought the mole was on her right shoulder.

He forged on. What was Patty wearing? Did she wear glasses? What colour were they? What was the number of her passport?

The number of her passport? You've got to be kidding, Dan, I thought. I bet you wouldn't know the number of your own passport, let alone that of your sibling's.

What was her social security number?

Amazingly, my mother, my extraordinary mother, rattled off her dead daughter's social security number.

Was she wearing a watch? What was the watch like? Mother described it, in words which I repeated to my interrogator.

'Uh-huh,' Dan said, 'a cheap watch.'

Inexpensive, I thought, might have been more tactful.

How did she get to the airport? Did she have a boyfriend? (No, no boyfriend.) Who was her doctor? His address? Who was her dentist? His address?

On the subject of dentists, Dan became draconian. I, personally, was to contact her dentist, to obtain Patty's complete original dental file and films, and to forward them to the RCMP in Halifax, marked 'Attention Swissair Investigation', with her name clearly on the front of the envelope.

'But, Dan,' I said reasonably, 'you already have that information. I understand that a Delta representative delivered it by hand several days ago.'

'I'm telling you what to do,' Dan growled. 'The original file and films.'

'How can I send them to you if they're already there?'

Dan wasn't listening. He repeated his instructions, word for word, as he read them off his computer.

We moved on. Was there anything about my sister that could possibly aid in the identification of her remains?

As it happened, there was.

'She had a plate and three screws in her right ankle,' I said. 'They were palpable, and just about visible to the eye.'

The plate and screws were the legacy of a broken ankle. Several years earlier, Patty's ankle had snapped during an ice-skating lesson in Central Park. She had hobbled back to Fifth Avenue, hailed a cab and somehow gotten home, refusing to seek medical attention until the following day. She was hospitalised for more than a week. When told that she needed an operation, she was incredulous and refused to consent. Only after the consequences were sketched out, did she capitulate.

Being off work proved therapeutic. Patty rested and grew slim and trim. The regular hours cleared her skin, vanquished her insomnia. She became good-humoured and contented, an advertisement for how wonderful life could be without working for an airline.

Then she went back to work.

In due course, the surgeon advised her to have the plate and screws removed – another operation with a six-week rehabilitation period. Predictably, Patty refused. 'Who has the time?' she said. I tried to persuade her to have it done. The plate could become infected, I argued.

She refused. My sister loathed being told what to do.

When at last Dan was finished, I scavenged for the day's *The New York Times*. In deference to Mother's sensibilities, I kept out of the lifts. From the newspaper, I learned that the flight data recorder had been recovered, and that the cockpit voice recorder was believed to be nearby. However, it was other information that caught my eye. It had to do with the state of the body parts that were being plucked from the sea. The medical examiner, Dr.

John Butt, was quoted: 'This is an issue involving, regrettably, fragmented remains. Nobody would be able to view these and find a point of identity.' If what they had was beyond recognition, Dan's moles and tattoos were not going to be enough for visual identification. However, as computerised data, it would contribute. A few years earlier, I had read a book about forensic pathology. The computer could throw up a list of potential matches for any body part, based on all available information. And in the end, there was DNA, the magic blueprint that had identified the bones of the long-murdered Romanoffs, the Tsar and his family.

Dan or no Dan, I had no doubts that Patty – or part of her anyway – would be identified.

TUESDAY, 8 SEPTEMBER

5:30 am. I got up, because the pressure of lying on my sore gut was unbearable. Inside my head, things were worse. Patty's fate appeared ghastly and unjust. What was she doing at the bottom of the sea?

I set myself the task of clearing the mess on the living room floor, the mountains of paper that had accumulated since I had done it five months earlier. There were stacks of newspapers and hundreds of loose scraps of newsprint, annotated in Mother's hand. There were notes scribbled on notepads, mail order catalogues, financial reports, medical newsletters, letters soliciting money. Everything was jumbled together in an untidy collage that stretched across the floor. Each item had to be examined, for fear of throwing out something important. And anything with a name on it had to be separated, for Mother was insistent that absolutely nothing could be placed in the refuse room with her name on it. As I worked away, a bit of newsprint fluttered out of a magazine, a poem by Emily Dickinson. I read it, then tucked it into my diary. By 9 am the job was done.

I telephoned the dentist, a gentle giant who had known Patty and me since we were children. 'I feel really, really bad about this,' he said, twice calling me Patricia, and twice apologising profusely. 'Call me Patricia as much as you like,' I said. 'It's an honour.'

As I expected, the dentist confirmed that he had given Patty's dental records to Delta. So I phoned the woman from Delta. I told her about Dan, and his fixation with the dental records that he already had. Thankfully, she said to leave it with her, and I wasn't to think about it again. However, the woman from Delta had other things on her mind. She wanted me to meet with her Personnel Benefits people. The idea made me cold. I did not want to sit down with people from Delta and talk about money.

There was another matter. Patty had kept a bag at the hotel in Geneva where she stayed on her layovers. The hotel had forwarded the bag to New York. Did we want it? I asked my mother, who had been listening to the calls. 'No,' she said emphatically. 'What terrible things we might find in there.'

I thought about it. What could be in the bag? A nightie, a swimsuit, cosmetics: the bits and pieces of Patty's life in Switzerland?

'No,' I said. 'Dispose of the bag and its contents.'

Later on, I was sorry that we had done this.

Adam, a cousin on my father's side of the family, was on the phone. I hadn't seen him in years, but I knew that he had taken Mother out to dinner the night before my arrival. He embarked on a convoluted tale. His wife had just come back from Chicago, and had brought a lot of pork with her, and would we be able to do him a favour?

It was a charming, quirky way of asking us to dinner.

My mother grabbed the phone. 'Don't you know that I'm Jewish?' she bantered. 'I don't eat pork. I may not even be able to eat the sauerkraut, I talk so much.' My mother was not Jewish, and she loved pork, and the gaiety of her chatter bordered on the manic. When she hung up, it faded swiftly.

In the newly tidied living room, Mother was talking about Patty. 'This last summer she was getting herself together, really changing. She was eating less quickly; she'd lost a few pounds. She was settling down and planning to retire.' As I listened, my mother cycled round and round the same themes. She was persuading herself through repetition, I realised. She was convincing herself that when Patty died, everything was on the up, that Patty was getting happier.

Suddenly, she interrupted herself. 'Go in there,' she ordered me, jerking her head towards the bedroom. 'Phone her up at once, and tell her to get over here immediately.'

In the lift I encountered Jesse, an elderly antiques dealer who lived in the building. The most beautiful flowers in the living room, the white orchids, were from him. He wrapped his arms around me. There were tears in his eyes. He was too overcome to speak. The other man in the lift watched with astonishment. We looked an unlikely couple.

I was going uptown to the office, the small space in the East Eighties from which my father ran his business. My father is gone, but the business has been taken over by my cousin, Henry, with the help of his father, my Uncle George.

On the way, my aching abdomen reminded me that I needed to eat. In a gourmet supermarket, I packed some salads into a carton. Then I sat at an outside table and munched. It seemed unspeakable to be eating alfalfa on Third Avenue with my sister dead off the coast of Canada.

In the office Henry and my uncle hugged me. My uncle asked at once if they had recovered Patty's body. 'Not that I know of,' I said. 'I don't think it's bodies that they're finding, anyway, so much as pieces.'

My uncle winced.

I busied myself with the paper-shredder, because I had lugged uptown a bag full of paper with names on it, the stuff that Mother wouldn't leave in the refuse room. I watched our family name, *Eberhart,* being sliced into ribbons, over and over again.

On the way back, I stopped for coffee. My mind was full of death – my father's and Patty's. But deaths come in threes. *Pan-pan-pan.* Who would be next? Like Patty, I had thought about Mother's dying, but not in terms of growing infirmity. Mother, I believed, was of too excitable a temperament, more likely to keel over in the street than waste away. However, I had invented something else to worry about: how Patty would react to Mother's dying.

They were close, too close. I could remember Patty's hysteria, when Mother lay sick with flu. 'You're not dying on me, are you?' she had snarled. Recently, more than ever, Mother had been the 'significant other' in Patty's life. *No boyfriend*, I had told Dan. Nor had there been one for years. In fact there had never been a serious boyfriend, not of the live-in, affianced variety. And my sister's friendships with women were, by my standards, very reserved.

We were like an old married couple, Mother had said. Patty called her last thing before a flight, and first thing upon her return. She couldn't get her hair done without consulting Mother first. She never bought anything, no matter how trivial, without showing it to Mother. And since Mother was always there to share the minutiae of her daily life, her death would have left a crater in Patty's life.

Now, as I sipped my coffee, the irony hit me hard. I had agonised over something that would never happen. The actuality was worse. Patty had died first; fate had subverted Nature.

I combed *The New York Times*, for the latest information about the crash. I read about the MD-11's built-in strategy against fires of electrical origin: a switch that cut a third of the plane's electrics at a time. I read that the flight data recorder had stopped, just as the plane had lost contact with air traffic control, six minutes before the crash. Then I pictured the pilots in the hot, smoky cockpit, registering the horror of full electrical failure. There was speculation to consider too: should the captain have attempted to land without dumping fuel, even though his plane was thirty metric tons overweight?

Finally I read more about the 'fragmented' remains. The next day's plan in Nova Scotia was to trawl for remains floating beneath the surface. The report was dated 7 September. So, while I had been walking, eating alfalfa and drinking coffee, people had been trawling with fish nets for my sister's body parts.

I stepped over the clutter in my mother's bathroom and climbed into the tub. As the water splashed onto my thighs, I remembered sharing the same tub with Patty. Members of a children's theatrical troupe, we had been acting in Shakespeare's *A Midsummer Night's Dream*. Night after night, Mother had washed the golden spray out of our hair, leaving a ring on the bath. At six years of age, I had played Cobweb, a fairy. Patty was Egeus, Hermia's uncooperative father, the blocking figure of classical comedy. Already, at ten years of age, my sister had been identified as someone out of tune with youth and festivity.

The bathroom door creaked open.

'I won't look,' Mother said. 'I just want to read you something.' She launched into some esoteric twaddle, the relevance of which escaped me.

'Mom,' I said tartly, 'I wish you would leave me alone while I'm having a bath.'

⚜

My mother had fallen asleep in her chair. Her glasses were on. *Newsweek* was open on her lap to the cover story about her daughter's death. Her breath rasped, seasoned by cigarettes. Her chest looked sunken beneath its red tee shirt. On the floor around her were the day's stack of condolence letters.

Suddenly she coughed with some violence, although it didn't wake her up. Had it been only five days earlier that I couldn't get her on the telephone? When I had imagined her collapsed, dead and alone, perhaps in that very chair?

The scent of flowers in the living room was heavy; the heat was appalling. My thighs were wet where they touched one another. The roar of traffic filtered up from Second Avenue. An airplane droned overhead. Why Patty? I wondered. Why not me? After all, I'm the one who was terrified of flying.

A memory flickered. Years earlier, we had taken off from Shannon airport in Ireland, Patty, myself, and my young son Richard. At once, the fear had taken me over.

'What's the matter?' Patty had asked.

I couldn't answer. I couldn't squeeze out the words.

'Now you're making me afraid,' she had said, turning white, having absorbed the terrible potency of my fear.

The memory made me feel stupid and selfish.

My mother slept on.

I struggled to remember the last time I had seen Patty. Was it in the room where I was now? Had she been sitting in the chair in which I was sitting? It came back slowly. I had been about to return to Ireland four, perhaps five, months before. Patty had brought salads from the Korean grocery. I flailed in my memory for details. Surely I should remember everything?

Pineapple, I thought. I had given her the piece of fresh pineapple out of my salad, because fresh pineapple makes me ill.

Pineapple? I wanted to remember everything about the last time I had seen my sister, and all I could come up with was pineapple? Why couldn't I do better than that? On what terms did we part? Our relationship had always been fraught. Was I glad to see her go that day? Was she angry with me? What did we say to one another?

I could not remember.

Only the last gift I had given her came into my mind. It was a tote bag with a Latin motto on it: EMPRIX NATA SUM. Born to Shop.

'Bye, bye, Miss American Pie.' The lyrics of Don McLean's ballad drifted through my head. The sense of my own inadequacy was overwhelming. I felt a tremendous need to become a better person. Why had I survived? Why wasn't it Patty who was sitting in the living room with our sleeping mother? A scrap of newspaper fluttered from my diary. I picked it up. It was the cutting I had retrieved that morning from the mess on the floor, the fragment of a poem by Emily Dickinson.

I read it again:

> 'Hope' is the thing with feathers –
> That perches in the soul –
> And sings the tune without the words –
> And never stops – at all –

WEDNESDAY, 9 SEPTEMBER

It was late morning. Aching with fatigue, I tried to nap, but lying down exacerbated my stomach troubles. So I wandered into the kitchen and seized a handful of jelly beans, bought with Patty in Connecticut. As I shovelled them into my mouth, the poster on the wall caught my eye. It was an old Air France poster with an Indian scene: a maharaja and maharani on the back of an elephant. It had been papered on to the wall forty years earlier, right after Mother had come back from an around-the-world tour. She had been gone for three-and-a-half months, leaving Patty, fifteen, and me, eleven, alone with our father.

The most interesting thing about the poster was the airplane in the background. The paintwork beneath had cracked, as soon as the poster was hung, and the crack ran through the tail section of the plane, breaking it away from its body. My father had stared at the divided plane. What did it mean? he had demanded. Was it unlucky? We had laughed uproariously at him.

'Funny as a crutch,' my father said.

I had laughed once Mother was safely back. But when she embarked on that around-the-world trip, I wept, because airplanes terrified me. In December 1960, two planes had collided over New York City, then crashed, one onto Staten Island, the other onto Brooklyn. These horrors had marked me. At eleven, I couldn't imagine how anyone could survive three-and-a-half months of going up and down in airplanes.

The phone rang. It was Frank, my husband; he had just landed at JFK. He had been met at the door of the aircraft by the fivesome of Delta officials. One of them had handed him a mobile phone. My number had already been dialled and the phone was ringing. All Frank had to do was say hello.

I was glad that he had come. I walked downtown to his hotel.

He had been given a large and comfortable suite, two rooms and a bath. He had brought with him a stack of letters, mostly condolences. Their warmth was bracing, extremely supportive. The last letter was from an acquaintance in Northern Ireland, who had asked me to contribute to a magazine she was editing. Inside was my essay, returned with a rejection slip. My subject, as it happened, had been death.

In the evening, we waited for my mother on a bench on First Avenue. Peering north into the city night, Frank spotted her, a shuffling old woman with a bum-bag. We ate in the greasy diner where Mother was now best friends with the Indian waiter.

Afterwards, Frank walked us home. On the street outside my mother's building, he kissed me goodnight. She complained. He should not, she said, demonstrate his affection in public.

⚜

The day's *The New York Times* contained a more complete transcript of the final exchanges between air traffic control and Swissair 111. I studied it, thinking of the pilots in their oxygen masks, struggling with heat and smoke. Not only did they have to fly the plane, absorbing information from instrumentation and air traffic control. They also had to tackle emergency checklists, while thinking about landing from cruising altitude with full fuel tanks in an unfamiliar airport. And they had to make decisions: quick decisions.

A single minute in the transcript stood out. The plane asked if it could dump fuel over land during its descent. The controller did not answer yes or no. Instead he asked, 'Are you able to take a turn back to the south or do you want to stay closer to the airport?' Now, it was the busy plane that did not reply directly. Eventually an answer came: 'OK, we are able for a left or right turn towards the south to dump.' And so the aircraft turned away from the airport, headed out to sea where it crashed ten minutes later.

When I went to sleep, I dreamed of another ending. I saw the plane descending riskily – too fast and too heavy with fuel – and I watched it crash-land at Halifax Airport, careering off the end of the runway. A few survivors scrambled from the flaming wreckage. Among them was Patty.

THURSDAY, 10 SEPTEMBER

Frank was waiting in the lobby of my mother's building. We were going downtown to see the lawyer who was handling Patty's estate. Since my mother regarded the subway as dangerous, I had promised to take the bus, but the bus was hot and crowded. After yet another night of stomach disorders, I was shaky. The lurching bus, the city smells and sweating passengers made things worse. After an hour we got off, hiking urgently westwards to arrive late. Although it was hot, my hands were like ice.

In a haze, I listened to details of assets, provisions, tax liability and schedules. The word 'compensation' was mentioned, with the prospect of lawsuits arising out of the crash.

It was unreal.

The lawyer sketched out the difficulties of proceeding without a death certificate. Then he produced an affidavit with my name on it, stating that I believed my sister to be dead since she had been on board Swissair 111, which had crashed without survivors. I watched my hand pick up a pen and sign, putting my name for the first time to the black-and-white of Patty's death. By now I was ice cold all over. Frank gave me his jacket. The lawyer disappeared to turn off the air-conditioning. Eventually, he suggested that we break for lunch. He took us to what seemed a vast, hollow chamber, but I didn't want to eat. It felt wrong to be eating, having just written away my sister's life.

Back in the lawyer's office, the magnitude of my sister's holdings began to dawn on me. I recalled her precise records, her stress at tax time, the pride with which she had drawn up her annual statement of net worth. Conservative, thrifty and obsessive, she had amassed a small empire. What use was it to her now? What real pleasure had it ever given her? Why hadn't she spent it on a nice apartment, on fulfilling her dreams?

In the taxi, I slumped in my seat. The dark, winding streets of downtown Manhattan pressed in. Then the cabbie pulled on to the East River Drive. At once the sky opened around us; the river glittered. It was a breath of air after the closeness of the lawyer's office, the tangle of affairs and figures.

❧

As arranged, we arrived at my cousin Adam's home together with Mother. A big doll's house dominated the hallway, a miniature of the brownstone itself, once a toy for the daughters of the house, and now a splendid conversation piece. We were taken on a tour of the real thing. Throughout, there was taste, order and comfort. In this house chairs were used for sitting and the kitchen table for eating. There was even a large, eccentric dog, and a garden for the dog to romp in. Right away, these remarkable people offered my husband a place to stay. The gesture touched me. Here was a man whom they hardly knew, a man unwelcome in his mother-in-law's apartment, and yet they were prepared to take him under their roof. A kind hand stretched for a moment over a divide. I was as happy as I could be under the circumstances.

Patty's name wasn't uttered. No one mentioned the plane crash.

I sat quietly eating the pork from Chicago off a beautiful china plate. My mother chattered with gusto. She was recollecting an occasion in Dublin in 1976. I had made dinner for her and

Frank. 'My daughter licked her plate,' Mother said. 'I asked Frank if he didn't think that disgusting, but he just grinned and licked his plate too.' My mother laughed merrily, then picked up the plate in front of her, and licked it clean.

'We like your mother,' Adam's wife said, as we were leaving. 'She's a pistol.'

We walked home along Second Avenue. Again, when my husband kissed me goodnight, Mother made her disapproval clear. I wanted to go back with Frank to his hotel. The prospect of his company, a shower, a television that could be watched, and a mattress that didn't sag, was enticing. What could be more normal than a husband and wife wanting to spend the night together? But I knew that my mother wouldn't like it, that it would make her angry.

In *The New York Times* I read that the boxer Jake La Motta had filed the first lawsuit against Swissair and Boeing, citing wiring problems, on behalf of his son, Joseph, and Joseph's companion, both of whom had perished in the crash. The damages sought were $50 million. My heart ached for Mr. La Motta. From the tabloids, I learned that he had lost his other son to cancer only a few months earlier. This swift, sharp lawsuit seemed to be a scream against the too cruel fate that had stolen both his sons. A lawsuit, at least, was action.

FRIDAY, 11 SEPTEMBER

I contemplated the difficult day ahead. A memorial service for the victims of Flight 111 was to be held in the middle of the day at the Avery Fischer Hall in Lincoln Center. The idea was unsettling. Lincoln Center was where we went to the opera, not to mourn.

The lawyer telephoned. A meeting had been set up with the

Personnel Benefits people from Delta. I scribbled down the agenda, then read it back, to make sure I had it right. My mother appeared at the doorway during the final stages of the call.

I hung up the phone.

Mother beaded her eyes on me. 'Ivy, you are not to regard your sister's death as a stroke of good luck.'

A stroke of good luck? How could she say such a thing?

'Patricia would know what I mean,' she said archly. 'It's the sort of thing that used to get her mad about you.' Feeling as if Mother had punched my sore gut, I walked away. She pursued me. 'It's all very well to be focused, but this is my show. I am in a black hole. I need to climb out. You should be helping me, not pushing me back in.'

Her show? Pushing her back into a black hole? 'That's unfair,' I said heatedly.

'You're not getting it. In my brain, I'm a very high-class woman. But I live like a pig. And I don't like it.' Then she sailed off towards the kitchen.

She's a pistol, Adam's wife had said. That she was, but it seemed unnecessary to shoot her bullets into me. And yet, offended as I was, there was no point in taking her seriously. Things were bad enough. Half-an-hour later, as we were heading for the lift, she said, 'It's a matter of perception. People don't perceive one another accurately. What I really said, Ivy, is that you've always been a lucky person.'

'Yes, Mother.'

Frank was downstairs. Because it was warm, we moved outside to wait for the woman from Delta.

A long, glossy limousine slid up. Mother, who is hostile to even the modest luxury of a taxi, began to sneer at this startling vehicle.

'Sssh,' I said. 'I suspect that's for us.'

It was. The woman from Delta was inside, together with

another woman, a senior vice-president from the airline. As my mother got into the car, she began to talk. A stream of words cascaded from her mouth, a flow so torrential that it required little in response beyond the occasional nod or yes. She didn't stop as we drove towards Fifth Avenue, then through Central Park. She talked as we climbed out of the car, as we were ushered into a waiting room with cookies, drinks and scores of boxes of tissues. It was as if every thought in my mother's brain came flying out turbo-charged, some of them vividly articulated, others impossible to follow.

As far as I was concerned, since she was eighty years old, and had just lost her daughter, she could talk as much as she liked, but I hadn't the energy to listen. Fortunately, the woman from Delta did. As the room filled with quiet, sorrowful people, Mother kept going. At one point I heard her announce, 'I don't know about you, but I'm having a good time.' I looked around, concerned that some might be offended, but all were caught up in their own heartache. It didn't surprise me that my mother had her own way of coping.

At last, we were taken into the front of the crowded concert hall. I opened the programme. The names of the dead leapt off the page. My eyes locked on to Patty's. What was Patty's name doing on a list of dead people? I read through the other names. There were so many of them. It was shocking to find the same surnames twice, three times, four times. Even five times! To have a single name on this terrible list was too much, but two surnames were repeated *five* times. How do you cope when five family members die without warning?

The service began with a blast of bagpipes. The pipers, from the New York Police Department, were tough, vigorous men, and they played with panache and a spark in their eyes. It was welcome and bracing.

When the service was over, we were surrounded by Delta personnel, friends of Patty's and acquaintances, and people who had never known her at all. Senior personnel had flown in from Atlanta, including the chairman and his wife. Their sympathy was prodigious, sharpened perhaps by an awareness of the lottery of death by plane crash.

By the time we left, I was drained. By chance, the black limousine took us home past Patty's apartment. It was unnerving to see the familiar door, untouched by what had happened.

Magnanimously, Mother said that Frank might come upstairs with us. The apartment was hot, on account of the antiquated air-conditioning that she refused to update. We needed something cold to drink. In the bottom of the fridge there was an ancient bottle of Asti Spumante, encrusted with food. I took it out and suggested that we drink it.

'No, no, no,' Mother said. 'Put it back. Clean it first, then put it back. If you want a drink, take something out of the closet.'

The closet was chock-a-block with old rums and bourbons, brandies, whiskies, and liqueurs, hardly the thing for a hot afternoon. So, after tackling the congealed food on the Asti with a knife and some Brillo, I headed out the door. I bought chilled sparkling wine, which we drank with some plum cake that Adam's wife had sent home with us. And Mother kept on talking, throughout the long afternoon.

After Frank left, Mother embarked on a detailed analysis of his shortcomings. I didn't listen. I wasn't interested.

That night, as I tried to sleep, Mother's words – about regarding Patty's death as a stroke of good luck – bounced around in my head. Bit by bit, I found a new way to look at them. Yes, they were a lashing out, intended to injure and demean. But so what? If my mother could say something so grim on such a day, then who could ever say anything worse? Just so, the experience could make me strong.

SATURDAY, 12 SEPTEMBER

Frank, who had not been inside my mother's apartment for decades, could not believe its aura of decay. 'Why won't she spend a few thousand dollars to make herself comfortable?' he asked.

I tried to explain. My mother had grown up poor. Born in 1918, she had survived the great flu epidemic during her first uncertain months. Her father, who joined the workforce at age twelve, had been salesman, seaman and train steward, among other occupations. During the 1920s, her family, which by then included a brother, Pierre, moved throughout the Greater New York area: from New Jersey to Yonkers to Queens, to the Bronx, Manhattan and Staten Island. A teenager during the Depression, she left school at sixteen to serve behind the counter at Barricini, a chain of chocolate shops. Having cut her teeth on privation and thrift, she was never able to live comfortably with prosperity.

Although marriage brought financial security, small sums remained her preferred currency. We wore second-hand school uniforms, as she stalked thrift shops and sales. For Mother, the bargain became the end in itself. It didn't matter whether or not the item was used. A good buy could be hoarded for a rainy day. And as the years rolled by, the bargains piled up, unused, in ever more mountainous quantities.

With her rigid ethic of waste-not-want-not, nothing was thrown out. If lettuce came in a plastic bag, that bag was washed, folded and stored. When butter was opened, the wrapper was scraped for the smidgen stuck to the paper. More than the meat was eaten on a chicken. Bones and giblets made soup, and every bit was force-fed to us, from the bone marrow to the pimpled rubber of boiled skin. And, over the years, a lot more than news-paper was retrieved from the building's refuse rooms.

Economy was my mother's friend, and conspicuous

consumption her foe. In a restaurant, she ordered modestly and brought home the sugar. When she travelled, she stayed in cheap hotels and brought back the soap. She never took a taxi, and if the apartment needed painting, she would do it herself. I pointed out to Frank that although the apartment was a wreck, there was plenty of paint: a dozen cans in the shower stall; another dozen in the hall; and more in a cupboard. There were drop cloths too, and turpentine and brushes galore, everything bought at knock-down prices. My mother saw no irony in the fact that the actual painting never happened.

As long as her ancient television provided the choice of two stations with fuzzy reception, why should she waste money on a cable? Besides, she would have to let someone in if the cable was to be installed. And didn't Frank know too well how my mother felt about people in her apartment?

We entered Patty's building, Frank, myself and Mother. The stairs creaked under our feet. One by one, I turned the keys in the heavy locks on her door. It swung open. My heart thudded. We were intruders; we were invading Patty's space. The air sighed and heaved. It belonged to her, even though she was gone. Her city handbag waited on the floor. There were clean dishes in the drainer, just-ironed shirts hanging over her bed. Her nightie was on the back of the bathroom door. I gripped it in my hand, smelled the scent of her body.

The disorder in the apartment was ferocious, not unlike Mother's. There were heaps of things everywhere, a clutter of excess. Mother was focused on finding some family pearls. 'They might be in the refrigerator,' she said.

The refrigerator? Was that because a burglar wouldn't look in the refrigerator?

'I'll find them,' I said. 'I'm good at finding things.'

Frank settled in the living room and began to sort through paper.

Like a thief, I prowled. The apartment would have to be emptied. I needed to assess the job. So I opened and closed everything. The sense of fullness was overwhelming. When I opened drawers, I could barely squeeze them shut again. Everywhere there was too much. Patty had travelled, and collected. Her apartment was a warehouse. Clearing it out was going to be hard. Her collection of wire chickens said it all. There were a dozen in a storage basket: enchanting, barnyard fantasies that she had bought in Central America. I had ten already in Ireland. There were others in Mother's apartment, and the house in Connecticut. No doubt Patty had given away even more. Just how many wire chickens had my sister bought?

As with wire chickens, so with everything else.

I despised myself for searching Patty's possessions and for judging her, for rooting through wardrobes and drawers, the intimate crevices of her existence. These were places I was never meant to be in, things I was never meant to see.

I found the family pearls and handed them to Mother. They hadn't been in the fridge after all.

SUNDAY, 13 SEPTEMBER

That night, my stomach problems deteriorated. My bowels turned themselves inside out eight times before morning. My face in the mirror was white.

Frank was flying home. We had lunch, then walked to his hotel, where he was collected for the airport. I wandered uptown with my throbbing abdomen. As I put in my key, the door of Mother's apartment flew open. She was furious. 'Where were you?' she screamed. 'I thought you were dead.' I calmed her down. Although I had been gone for only a couple of hours, I could relate to her anxiety.

We went out immediately. As soon as the lift door closed, it jammed. Never had either of us been stuck in an lift. 'Sit down,' I said. 'Try to keep calm.' My mother ignored me, stalking around the confined space, raging at the nameless incompetents who had allowed such a thing to occur.

After a few minutes, the lift door opened.

The experience could not have lasted for long: five minutes at most. There was a bench to sit on, and an emergency bell to ring. The lights did not go out. We had each other, and nobody else. The experience was not remotely like the twenty minutes of crisis on Swissair 111. It was light years away from the last six minutes in a plunging tube of people who were about to die.

And yet it had been extremely unpleasant.

MONDAY, 14 SEPTEMBER

I was ready for the meeting with the Personnel Benefits man from Delta. I was wearing a black skirt with a dark blouse that had belonged to Patty. As I stood at the door, Mother complained. 'You look too dressed up. What will they think of you?'

Too dressed up? Was she serious?

She was. I remembered that my mother never liked us to look grown up.

'I don't care *what* they think,' I said to her. 'Besides, who could consider me dressed up in these shoes?'

It was the right thing to say. Mother has always detested my comfortable footwear. 'Terrible shoes,' she muttered.

I took the subway to Times Square, then walked south down Eighth Avenue. My last time on Eigth Avenue had been with

Patty. She had bought a huge potted plant with spiky fronds. We had lugged the clay pot uptown with other parcels, for once a humorous monument to excess acquisition. Now I was looking for a tall building, but I wasn't thinking straight. At one point I walked back, half-a-dozen blocks, before retracing my steps. Eventually, I arrived, ten minutes early.

I dawdled on a bench outside, not wanting to go in.

The lift powered 51 storeys upwards (one for each year of my sister's life) with a speed and thrust that did nothing for my battered stomach. I was a minute before the appointed time. My lawyer's face peered outwards through the glass door.

The offices were vast and empty. There was a receptionist, luxurious fittings and little else.

Waiting with the lawyer were the woman from Delta and the Personnel Benefits man. A dark, stubby individual, he had nondescript features except for his eyes: these bulged through thick corrective lenses.

The four of us settled in the plush board room. Light streamed through the windows. The city stretched below in a spectacular panorama.

The comforts belied our business.

I was handed one cheque after another: Patty's final paycheque and a couple of insurance cheques. She had been paid, I registered, for her final flight. And a flight or two afterwards. Blood money, I thought. Straight swap. Patty's life for stacks of money. All the luxury in the world could not sweeten the horror of what those cheques represented. I noticed that my name had been spelled wrongly on one of them. It seemed in keeping with the occasion.

Integrity lay at the root of my sister's personality. She was a dedicated employee who had gone to work with her employer's interests always in mind. In turn, she expected her company to play fair, to treat her with respect and trust. She became angry when she believed this wasn't happening. Meticulously exact in her own financial dealings, she expected the same in return. She knew what her entitlements were, and didn't doubt that she had earned them. Thirty years' service would be rewarded with airline passes for the rest of her life. As far as Patty was concerned, she could not retire until she had earned that perquisite. She had talked about it a lot. At the time of her death, she was three weeks short of her thirty years.

I understood my sister's thinking. I knew that she would expect me to ask for her entitlements, for what she had earned. She would have felt that I had let her down if I didn't. So I asked the Personnel Benefits man for Patty's travel benefits to be transferred to me. It was the one moment during the meeting where my emotions were not controlled. I heard my voice wobble.

My request was refused.

When the meeting was over, my lawyer asked if he could stay with me in the board room for a brief conference. This was agreed, and we talked for twenty minutes. When we came out, both the Personnel Benefits man and the woman from Delta were waiting. Apparently, they could not leave us in those vast empty offices under the supervision of the receptionist only.

The lawyer and I descended the 51 floors. The cheques were in my handbag. My ability to make decisions was at a low. 'What am I supposed to do with these cheques?' I asked.

'Treasury Bills,' the lawyer said.

I had no idea what Treasury Bills were and I didn't much care. I simply wanted to get the cheques out of my handbag. The lawyer studied me with the mildest trace of impatience. 'Would you like me to come with you?'

The question helped.

'No.'

I headed uptown towards the Allied Irish Bank, photocopying the cheques on the way, treble-checking to make sure that I hadn't left them behind on the machine.

Inside the bank a lovely Irish woman ushered me into her cubicle. I stated briefly what had happened, and what I wanted. The young woman fussed over me, explaining everything in words of one syllable. When I fretted about the cheque made out to Barrister instead of Bannister, she told me to sign so it could be read either way. Her warm efficiency soothed me. My feeling of oppression lifted slightly.

TUESDAY, 15 SEPTEMBER

10 am. I was waiting on the street outside my sister's building for the assessor, who was to value Patty's possessions so they might be taxed. It was a beautiful morning. A bird trilled in the branches overhead. The entire street was washed in green, as sunlight filtered through the foliage. In front of a fashionable café, chairs and tables were being set out. A Lincoln Continental purred from a garage. It was idyllic, the Manhattan of dreams, only I was rigid with cold. I was about to let a strange man into my sister's home to reduce her belongings to a figure.

The lovely street had always been a part of my life. My father had managed the building when we were children, and Patty and I had longed to live there. 'In this world there are only two tragedies,' Oscar Wilde wrote. 'One is not getting what you want, and the other is getting it.' The once longed-for apartment had not brought Patty happiness. For more than twenty years she had been talking about moving.

I was halfway up the stairs with the assessor when I heard the front doorbell. The lawyer. I handed Patty's keys to the stranger, explaining the intricacies of her locks, and descended again to the street door. My sister would have been appalled by this cavalier treatment of her keys, my flagrant breach of security.

Upstairs, as the two men exchanged a few words, I stalked through the apartment. Suddenly a pair of panties leered at me from the bedroom floor. I grabbed them, balled them into invisibility in my palm. How had I missed them that first night?

The assessor took out his pencil and notebook. The lawyer settled at the alcove table with Patty's papers. I opened a plastic sack and started heaping her knitwear into it. In moments it was full, as big as a body bag, only more bulging.

The two men were as quiet as mice.

I snapped open another sack.

The assessor asked about Patty's jewellery. I showed him what I had found: a wooden chest with twenty drawers; a box in her bedroom with earrings and watches. He opened and closed the many compartments. 'But none of this is real,' he said.

None of this is real.

The words hung in the air, more potent than he knew. I understood what he meant: that the bright mountain of ornaments owed nothing to precious metals or stones. But, as far as I was concerned, everything was unreal, from our presence in Patty's apartment back to the driving force of her horrible death.

None of this is real.

The assessor's face hovered in front of me. I wanted to help him complete his job. But what could I say? That I knew nothing about Patty's jewellery beyond the fact that she rarely wore any: not the bright and curious creations in front of us, nor the 'real' items he was looking for? That she was too modest to spend large amounts of money on herself? That she saw diamonds as connected to marriage, a happy state that had persistently eluded her?

None of this is real.

In the end, I stammered out what I could: the fact that Patty had lived in two other apartments in the same building, and that the first had been burgled repeatedly, which had discouraged her from acquiring costly jewellery. The assessor nodded, and turned instead to the china and crystal.

My mother arrived. She retreated to Patty's bedroom. There, in slow motion, she removed things from the wardrobe and put them on the bed. She ground to a halt over a ruffled blouse. With care, she matched it to a frilly skirt. She held them up together, a fluttering confection of baby blue. 'Tell me,' she said, 'how could a woman who wore clothes like these die in an airplane crash?'

The lawyer passed me a note in my sister's writing. 'Ivy,' it read, 'the policy is in the green folder on the bottom shelf to the left of the fireplace.' In thirty seconds, I was holding an insurance policy. I handed it to the lawyer, briefly touching his arm. He recoiled, whether from the chill of my fingertips, or my friendliness, I was not sure.

⁂

The lawyer was ready to go. As he picked up his briefcase, it scratched the table. Embarrassed, he began to apologise. The sight of the slight damage to the table sucked me into the past.

It was fifteen years earlier. Richard was three or four. Patty had suggested that our little family stay in her place, while she decamped to the parents' apartment. It was a generous, but loaded offer. We arrived to a list of house rules that ran to ten pages. Among the instructions were details about the dishes we were to use. Richard was to drink only out of a large paper cup, which was to be washed and reused throughout our stay. Soon, the cup became worse for wear. One morning, it collapsed in our little darling's hand, spraying grape juice, which spattered the fringe of a Persian carpet, adding a mild violet hue to a few inches.

When Patty arrived, the trivial stain was pointed out. She erupted. No short, sharp tirade, her fury took root, festering throughout the day. There was nothing that she didn't call us, no limit to our criminality for allowing such a thing to happen. As we walked up Madison Avenue with the stroller, stony silence prevailed in the face of Patty's angry monologue. By now, it was levelled solely at me: 'You have a husband,' she bellowed. 'You have a baby. All I have is my possessions. My possessions are everything to me. Everything!'

The glories of the Metropolitan Museum of Art did nothing to calm her. That night, battered by her fury, Frank dragged the carpet into the bathroom and washed its fringe. The violet came away, together with the dirt. Painstakingly, I dried it with a hairdryer. The result was fluffy and appealing. The next day, when the sparkling fringe was pointed out, Patty was not impressed. We should have done the other end of the carpet too, she complained, so that it would not look unwashed.

Back in the present, the lawyer was still apologising, fingering the scratch on the table. 'Don't be silly,' I said. 'I'm sure it was there already.'

⚜

In the afternoon, I returned to Patty's apartment alone. Thankfully, Mother did not object. Perhaps the few clothes she had left on the bed had convinced her that she was not up to the terrible task of dismantling her daughter's life. I turned at once to Patty's underwear, a straightforward matter, I imagined, of throwing everything out.

I opened the top drawer of a six-drawer chest. Panties eased up, freed from the stuffed confines. On top was the much-washed cotton that she wore, every pair perfectly folded. As I scooped them into the refuse sack, I could see her folding them obsessively, forcing away every whisper of wrinkle. A feeling of embarrassment washed over me. What was I doing with Patty's underwear? How would I feel about someone pawing through mine? Beneath the cotton were pristine French mini-knickers, half-a-dozen pairs. I had given them to her ten years earlier. Why hadn't she worn them? I ploughed deeper. Fruit-of-the-Looms. Unopened. Several packs.

I moved down a drawer. Inside, nylon bloomers trimmed with machine lace. The word *pettipants* floated into my head. We had worn them under our school uniforms forty years earlier. Under the pettipants, some ancient satin step-ins. Were they Mother's? Grandmother's? Why did she still have them? I stuffed them into the sack.

The second drawer was nearly empty. A small box rattled. Inside, still in its wrappings, was an Arcing Spring Diaphragm. I flipped the box over. *August 89* was printed on the bottom. Unused, unopened, ten years past its sell-by date. My heart squeezed. This intimate artefact spoke volumes: I felt the full force of Patty's yearning to fall in love, to connect, to have children. And the terrible tragedy of her failure.

Tears flowing, I ploughed on. A drawer of little bras. They were tiny. Could she possibly have been so small? I held up a wisp of candy-striped gossamer, dotted with Playboy Bunnies. The garment was almost asexual. Where had it come from? Had she ever worn it? Into the refuse it went, together with petticoats, as old as the hills. I snapped an elastic. It gave, perished. I pulled another. The same. Why hadn't she thrown these things out?

The apartment around me was silent, a deep quiet pierced by the occasional bleep from a gadget in the living room. The bleeps were meant to be offensive to cockroaches. I had never seen a roach in Patty's apartment. I dragged the refuse sack, heavy with intimate apparel, past the bleeper. If the bleeps were offensive to roaches, what might they do to humans?

From infancy, Patty had loathed creepy-crawlies. 'Bugs! Eugh!' I heard her say, and then I saw her before me, a child with a look of comic distaste stamped across her face, standing in the middle of our father's vegetable garden in Connecticut.

That night, as I struggled to sleep with my sour stomach, memories tumbled around me, memories that only Patty could share. I saw us sleeping together, children in flannel nighties that Mother had made. I saw us trudging through a storm to school with Father – who didn't accept that schools should be closed on account of a little snow. I saw us in ecstasy over a gingerbread house that Grandfather had brought for Christmas. On and on they went, pictures unscrolling, the lost images of childhood.

Then another memory pressed in, of two young women in a rented car on the ring road around Paris. We were young and careless. Having toured through Brittany and Normandy, we had just left our parents at Cherbourg; they were returning to New York on the *Queen Elizabeth II*. Now, Patty and I were bound for the cathedral at St. Denis, to see where the bones of the Kings and Queens of France had been dug up during the French Revolution. One last sight before our respective flights home to New York and Dublin. Unfazed by the unfamiliar road, we were driving at speed in an outside lane, when another car suddenly loomed ahead, approaching us head-on. Time suspended itself, as we contemplated what was about to happen. At the last moment, we were saved. Not by any action of our own – we were in the wrong – but because the other car pulled over.

Had the Fates struck then, I thought – had we both died in a cataclysm of crumpled steel, smashed glass and blood on that French road – that long-past disaster would have pre-empted this nightmare of the present.

WEDNESDAY, 16 SEPTEMBER

Iwas busy in Patty's living room when her telephone rang. It was the woman from Delta. She explained that Dr. John Butt, the Chief Medical Examiner of Nova Scotia, wanted to speak with me. She, herself, she added, would be joining me shortly to provide her support. I knew from reading *The New York Times* that Dr. Butt was the coroner in charge of identifying the remains from Flight 111. If he wanted to speak with me, there could only be one reason: Patty's remains had been identified.

I also knew that I wanted to be alone with Dr. Butt's voice. Of course I would have been glad of my husband's company. But Patty would not have wanted anyone from the airline management in her apartment. So I insisted on staying alone. Patty was dead. Nothing Dr. Butt could tell me would change that. Besides, knowledge is enabling. It brings boundaries to the horrors churned up by the imagination. Dr. Butt's words would throw light on the way that Patty had died. The closer I could get, the better I could come to terms with it.

The newspapers had provided a framework of information: I understood that recovery was slow, and remains fragmentary. Nonetheless, I had never doubted that Patty would be identified with relative ease. I pictured her as busy at the time of impact, preparing for an emergency landing. I did not see her strapping herself in. And so, when it was over, what was left of her unconstrained body would have surfaced quickly, unattached to wreckage on the ocean bed. The way I had imagined it, I assumed there would be enough to identify, with no seat belt to cut her body in half.

While waiting for the phone, I kept busy, filling sacks with her lifeless clothing.

Dr. Butt's voice was deep and musical. As I listened, I was sitting in Patty's tiny kitchen, my bottom on the steps that doubled as her chair. I was sitting exactly as she had sat herself when talking on the phone. I scribbled Dr. Butt's words on to the same notepad where she had written, exactly two weeks earlier, a few hours before her death. I sipped cold coffee, as he described what had been my sister. There was a torso with part of a head. Portions of her upper legs were still attached, more of one than the other. Positive identification had been made through dental information.

Patty's distinctive teeth gleamed.

'Arms?' I asked quietly.

'No.'

Part of a head, he had said.

'Were there any identifiable features?'

'No!' Dr. Butt's response was hasty. Too hasty. Then, as if realising that he had betrayed some horror, Dr. Butt elaborated. 'A dreadful impact,' he said. 'Tremendous dismemberment and fragmentation ... an accumulation from the surface that continued to come up for several days.'

I felt sorry for Dr. Butt. I had already done a quick sum. There had been 229 people on the plane, and an unborn child. At a modest average weight of 130 pounds per person, that meant 30,000 pounds of flesh – 15 tons of smashed-up people to recover, sort, catalogue, attempt to reconcile.

I asked when Patty had been recovered. 'Early on,' he replied. 'The pathological description of her parts was made at 9 pm on the evening of September 4th.' Two days after the crash. I hadn't even left Ireland.

In my head, I visualised boats on a heaving sea, nosing through wreckage. Grappling hooks struggled with fuel-coated husks of humanity. I saw living hands hoisting the fragmented dead over the side. On a slippery deck I pictured my sister's exposed body: glistening wet; raw-edged; armless; mostly headless and legless. Yet even as Dr. Butt spoke, I understood how relatively intact Patty's remains were. A torso, with part of a head and at least one tooth to identify her: completion itself by comparison with the pieces of flesh that divers were plucking off the seabed.

Dr. Butt sang the praises of the young men who were devoting themselves to this work. I thought of them: divers in pairs, deep in the eerie underworld, 180 feet beneath the surface. I saw them grope in the weird lighting for the bits and pieces of the dead mangled with wreckage, the divers themselves at risk of their air hoses being pierced, of getting snagged in the debris. I pictured the mesh bags rising towards the surface, their gruesome contents streaming. Such images to carry with them for the rest of their lives!

Dr. Butt said the recovery work would continue. They would be trying to 'fast track' identification of bits and pieces. It was possible that further remains would be associated with my sister. I listened until he was done. Then I asked Dr. Butt how old he was. When he replied that he was sixty-four, I was inexplicably relieved. For some reason, I wouldn't have liked it if he had said he was thirty.

After hanging up, I decided to eat something, to soothe the burning in my gut. In Patty's fridge, there was yoghurt. I took a spoon out of the drainer, a spoon that she had used and washed on the day of her death. Then I ate her yoghurt. As I swallowed, I thought of her tongue, and the teeth that had identified her, touching the spoon.

On the way back to Mother's apartment, I bought a loose-leaf notebook for my growing collection of documents and notes. Walking up from First Avenue, I was confronted by a startling sight. Emerging from the service entrance of my mother's apartment building were two men rolling a trolley with what was clearly a blanket-wrapped cadaver strapped onto it. Carefully they lifted the pencil-thin corpse into the back of a waiting hearse.

From the awning of my mother's building, one of the door-men watched with a grim face.

'Who is it?' I asked.

A young woman, I was told. She had lived on the ninth floor, and had died in her apartment of breast cancer.

❦

In the living room I told my mother about the conversation with Dr. Butt. I explained that there were no whole bodies, that only ten of the dead had been identified, and that there were unlikely to be further identifications for a while.

'So we're one of the lucky ones,' she said.

At her request, I described Patty's remains, exactly as Dr. Butt had described them to me.

'Venus de Milo,' my mother said.

The confirmation of unthinkable dismemberment brought its own consolation. An impact violent enough to smash off part of a head and amputate limbs could not have been felt. The lives of the victims had been switched off like lights. But the question of mental suffering was another issue. It stayed with me, inhabiting my mind.

That night, as I stalked the halls of my mother's building, searching for *The New York Times*, I made my way to the ninth floor. I found the door behind which the young woman had died of breast cancer. I stood outside and prayed for her and for Patty, tears flowing for the two dead women, both of them, in a way, complete strangers to me.

THURSDAY, 17 SEPTEMBER

Early morning. I was downing some orange juice when Mother appeared. 'Patricia was on a fast track to nowhere,' she growled. 'There was nothing ahead for her but suffering.' I was surprised. Since Patty's death, Mother had taken to speaking about her in radiant terms. Even her name had evolved, becoming Patricia, instead of Pat or Patty. So I changed the subject. A couple of days before the accident, I had discovered autumn crocuses in a Dublin garden centre. These Buddha-shaped bulbs would flower, apparently, on a sunny windowsill without water or soil. So I told Mother about the two that I had left in my kitchen. The story diverted her, and soon she was boiling water for instant oatmeal.

I left the apartment with only the juice in my stomach. Eating was too painful first thing in the morning. Besides, there was no fresh, wholesome food in the apartment. Nor was there coffee, or a mug to put it in, or an uncluttered table to sit at while drinking it.

On my way to Patty's, I stopped at the unloading bay of a supermarket. I explained to the man in charge why I needed boxes. At once, he heaped my arms with cardboard, and promised to save his best empties for me every morning. Then, I headed in the heat towards Lexington Avenue, balancing my tower of cardboard. Inadvertently, I jostled a not-so-youthful woman.

Under the canopy of her purple hat, she was not amused. 'What's the matter,' she snarled, 'are you moving?'

'No,' I replied pleasantly. 'My sister just died in an airplane crash, and I'm cleaning out her apartment.'

'Sorry I asked.' And she melted away in a disturbed purple fug.

The breakfast problem was solved on Lexington Avenue with a takeaway bagel and café-au-lait, a bland offering to consume bit by bit as I tackled Patty's apartment. I had settled on an approach to the colossal task. Everything would be sorted into four categories: for Ireland; for Mother; for the church; and for the rubbish. This last category presented difficulties of its own, because Patty, like Mother, had intense views about the disposal of refuse. Both were taken over by the idea that people were waiting to pounce on their rubbish with intent. These people, they feared, would gain information which would then be wielded to devastating effect.

Patty did not like anyone to know anything about her. None of her colleagues, let alone her neighbours, knew much about her personal tastes or finances. She met even the most casual questions with suspicious discretion. Not only would she not leave an addressed envelope in her refuse room. She would not leave anything there at all that might be associated with her: not a leaflet from the airline industry, or a financial report, or even a shopping bag that might connect her with Connecticut, where my mother owned a house. Patty had a shredder that she used heavily, disposing of the shreddings themselves with elaborate care. She had two stockbrokers, so that neither would be too familiar with her portfolio; and she received mail at four different addresses (not counting the airport), so that no snoopy postman would become too familiar with her correspondence.

So, at first, I tried not to leave personal items in her basement refuse room. Instead I wandered down Madison Avenue, dropping the contents of her two medicine cabinets, handful by handful, into the garbage cans of New York. Block by block, my sack grew lighter, shedding laxatives and eye washes, nasal decongestants, depilatory, mascaras and lipsticks. Ten blocks south, and I found myself outside a posh restaurant. A van was making a delivery. 'Live Longer, Eat Fish' it said on the side of the van. Fish, I thought sardonically; she should have eaten fish. Then a darker thought wriggled past: that the fish were probably eating her, gulping down those bits and pieces that were still loose in the sea.

I cut across to Fifth Avenue, and walked back uptown, a flick of my wrist at every garbage can, followed by a rattling clunk, as the vials and tubes banged against the metal. From time to time, I looked over my shoulder. There are laws against the improper use of the garbage cans of New York, and I was breaking them. What if I was dragged off to a police station with a handful of half-used unguents as evidence against me? The business of flinging Patty's belongings into such smelly repositories sickened me. It seemed a terrible fate for things she had touched. I felt I was disposing of part of her own self. In a way, I was finishing off the job that the airplane crash had begun.

Grimly, I mounted the stairs back into her apartment. There was a formidable double wardrobe in her foyer, bursting with clothes, shelves, and extra hanging space created by rails attached to both sides of the door. On the floor were dozens of shoe boxes. On my knees, I pulled them out. Each box was stacked precisely, each tower of boxes aligned. Inside the shoes lay end to end, a shoe tree in every shoe. Like a demon, I hauled out box after box, yanked out the trees, dumped the shoes in my sack, smashed the boxes for disposal in the refuse room. Soon, I was back out on Madison Avenue with a freshly bulging sack.

Pair after pair, they flew into the refuse: shoes that I recognised and shoes that I didn't. Four pairs of patent leather flats that Patty had worn as a teenager, mixing colours – one pink foot, one blue foot. Party shoes, with heels so wobbly that I couldn't imagine her wearing them; shoes so old that the pressure from the trees had cracked the toes. Relentlessly, I dumped them, overwhelmed by their femininity, dozens of little shoes, daintily bowed, with tiny heels and strappy backs, every pair a world apart from the walking shoes that I favour myself.

And so many of them!

With my shrinking sack, I plodded onwards, no doubt a very peculiar sight. Fortunately, New York is full of strange people, and nobody took much notice of me, a comparatively well-dressed bag lady-in-reverse: throwing it out instead of gathering it in. It was the 17th of September. The day before I had planned to arrive in America. I should have been back in Dublin, packing. The last time that I had talked to Patty, she had been full of plans about where we would be eating and shopping. And what if she had told me that I'd be dragging myself down Madison Avenue, scattering her shoes? How we would have laughed.

A phrase came into my head: *souls on board.* Twice, the air traffic controller had asked the troubled aircraft: *How many souls on board?* Neither pilot had answered, caught up, as they were, by more pressing demands. *Souls on board.* The phrase wouldn't go away, not even as Patty's shoes sank in the trash. *Souls on board.* A commonplace, perhaps, to air traffic controllers, but, to me, a statement that the fate of those passengers and crew was deter-mined, as if already they were souls, throughout those last terri-ble minutes.

By the time I got to bed, it was 11 pm. When Patty died, I had been reading *Captain Corelli's Mandolin* by Louis de Bernières. I picked it up again. For a few minutes I was in another world, a magical world that was a welcome escape. Soon I dropped off. I was in the deepest of sleeps when I felt my mother shaking me. 'Wake up, Ivy! Wake up!'

'What is it? What's happened now?' I grabbed my watch. It was 2 am.

My mother was clutching a newspaper in her hand. 'Autumn crocuses are poisonous,' she cried. 'Poisonous, Ivy, do you hear me? Whatever you do, don't leave them around. It's all in this article.'

'Couldn't it have waited until tomorrow?'

'Absolutely not. Just get rid of them. Immediately.'

I lay on the bed, my sick stomach throbbing. I did not get back to sleep.

FRIDAY, 18 SEPTEMBER

In the hot, lumbering bus with Mother, my sore insides grew more and more distressed. When at last we arrived at the lawyer's, I disappeared to the toilet, where a foul sea of sludgy strings exploded from my body.

In his courteous, helpful way, the lawyer went over various points relating to Patty's affairs. But the final item on his agenda filled me with dismay. It had to do with lawsuits. The lawyer elaborated with clarity. When an airplane crashes, he said, there is always a reason. Where there is a reason, there is negligence. Where there is negligence, somebody can be held accountable for the awful thing that has happened. And because there is accountability, someone can be forced to pay out compensation. In an air crash, there are plenty of possible defendants: the airline; the manufacturer of the plane; the company who maintained it; and so on. To take on these defendants efficiently, there were dedicated law firms whose business focused on airplane accidents.

I shuddered.

The lawyer continued. He knew of four firms that were active in the field. Swissair 111 represented 229 potential cases. Although each case might be filed separately, the courts would try to consolidate them into a single action, a class action. Therefore, each firm of attorneys would be keen to sign as many clients as possible, to give themselves leverage as things progressed.

An image of sharks came into my mind. Opportunists gathering.

I was grateful to our lawyer for telling it like it was. He wanted us to meet with all four firms, before I returned to Ireland. The idea that Patty's death had become a commodity for lawyers to bargain over was unsettling.

'I earned it,' I heard her say roughly.

Blood money.

In the meantime, Mother was getting excited. From what the lawyer was saying, it appeared that Patty's life, as a single woman in her early fifties, would be valued less highly than that of a married man in his thirties with two young children. My mother was outraged. 'A life is a life,' she insisted. 'My daughter's life was worth exactly the same as anyone else's on that plane.'

The meeting ended. Fortunately, the lawyer insisted that we lunch in a diner so modest that Mother didn't baulk. Eating meat loaf with mashed potatoes soothed my knotted insides.

When we got home, a lavish bunch of roses was waiting with the doormen. For Mother, I assumed.

The phone was ringing as we entered the apartment. Dumping the flowers, I raced for the bedroom, but didn't make it.

'Oh dear,' Mother said. 'That one must have been for my screen test.'

I found a vase for the roses. Then I noticed the accompanying card. It was addressed to me. Inside it read, 'Love from Frank.' I looked in amazement from card to flowers. My husband was not the floral type. Why had he sent roses?

As usual, Mother fell asleep in the living room. She looked old and unwell. Although the lawyers would argue that Patty had no dependents, it wasn't true. This ailing, elderly woman was Patty's dependent. *Like an old married couple.* Patty was my mother's link with the world. She was Mother's only companion: her dinner, movie and opera date; her partner to quarrel with. She was also Mother's chauffeur and confidante. Together, they had endured the trauma of my father's illness. Without Patty, Mother would be alone.

I poured myself a glass of wine, then sat down, kicking some newspaper cuttings out of my way. The mess depressed me. I was sick of the junk everywhere, the bulging plastic bags, the ragged army of overflowing boxes. What was the point of this plague of possessions, an asphyxiating clutter that interfered with living? I wanted to wash in the shower, sit at a table, hang up my clothes.

In the kitchen, there was a chest of drawers. The middle drawer was wadded with used Christmas wrappings, once folded for recycling, now brittle with age. I had asked Mother if I might clear the drawer. Half asleep, she had agreed. It seemed prudent to get on with it before she woke up, so I excavated the layers of faded wrappings and tarnished tinsel. Carting them out, I bypassed the nearby rubbish room. I didn't want Mother bringing it all back in again. In the bottom of the drawer was a squashed layer of the hats and scarves we had worn as children. I threw them out. Beneath them, wrapped in tissue paper, was a sterling silver frame.

The reporter from *People* had sent us enlargements of Patty's photograph. I slotted one into the silver frame. I cleared a few inches on a counter, so that my mother would see the photo each time she entered the kitchen. Then I arranged my underwear, shirts and trousers in the empty drawer. Dressing in the kitchen would be the height of convenience compared with rooting through my suitcase.

The phone rang. It was Frank.

I thanked him for the roses. 'But why did you send them?'

'What date is it today?'

'I don't know.'

'The 18th.'

'So?'

There was silence in Dublin. What was so special about the 18th of September? Then I remembered. It was our anniversary. And it was not only our anniversary, it was also Frank's birthday. 'Well, thank you,' I said, 'and Happy Birthday.' Suddenly, the arrival of the bundle of red roses seemed miraculous.

My mother stirred in the living room. Her eyes fluttered open, and she smiled. 'When I woke up just now, I wasn't thinking of Patricia,' she said. 'And then, I thought, maybe she's just gone out to get her hair bleached. Only she doesn't have any hair.'

SATURDAY, 19 SEPTEMBER

'Remember,' Mother said as I went out the door, 'your sister was as successful a metazoan as you or I. Only an accident interceded.'

In Patty's apartment, I turned again to the double wardrobe in the foyer. The space was packed with clothes, each outfit sandwiched between its neighbours. I could see only a sliver of each garment. Browsing was impossible, because nothing could be moved more than an inch. The back row of clothes was invisible and inaccessible without removing the stuff in the front. Patty always said that she never wore anything except her uniform or her nightie. Now I understood why. It was too much trouble to find anything.

I loaded one sack after another. Every item was immaculate; many were covered in protective plastic. But I couldn't help but think about how odd almost everything would have looked on her. What I wanted to find in that wardrobe were smart, well-cut clothes, clothes that would have enhanced Patty's slight physique. Instead, there were corduroy prints and floral poly-esters, a mishmash of long-past fads. Where was the simple sophistication that would have reflected Patty's travels and her smart financial mind? Why was she clogging up her life with masses of unfashionable garments that she never wore? Why couldn't she let go?

After a while, I penetrated to the back layer in the wardrobe. There I discovered a blue polka-dotted mini-dress, crotch-length, shapeless and diaphanous. It was one of a pair that my mother had made, thirty years earlier; and the only reason that I remem-bered was a funny photo taken of us wearing them in Florida. We had looked so different in the identical dresses! In spite of her bare thighs, Patty remained prim, her feet neatly shod and her hands clasped in a ladylike manner. She had pinned her collar closed with a bow. By contrast, I was all over the beach, loose-limbed and broad-shouldered, my throat bare. My shoes were in my hand and a naked foot pawed at the sand.

Why had Patty saved this ridiculous dress? I couldn't remem-ber what I had done with mine. I had probably chucked it out at college.

But no, I hadn't. Twenty garments down, I found the match. For thirty years, Patty had preserved both. Did she really believe that we would wear them again? That, in our sixties, we would frolic in polka-dotted mini-dresses?

In *The New York Times* I read about the wreckage of Flight 111 as it appeared on the sea-bed. Sonar readings had suggested a fuselage lying in a couple of pieces, possibly with bodies strapped inside. Inspection revealed a different story. One 'piece' turned out to be a large rock. The others proved to be heaps of pieces, jumbled on top of one another, everything small. And there were no bodies, only body parts, tangled up in the wreckage. '...[W]hat's left of human remains is hard to identify sometimes because it almost looks like cushions,' a diver was quoted as saying.

The wreckage was concentrated over an area smaller than the aircraft itself. This pattern confirmed that the plane had hit the sea at 'a sharp, almost perpendicular, angle' and 'at a tremendous rate of speed'.

⚜

In the evening, I raised the subject of Pierre with my mother. Pierre was her only sibling, and he had been dead for a quarter of a century. My memories of him were hazy: a slender figure, dark-haired with a charming smile, gleaming teeth and a moustache. When I was a child, he had disappeared from our lives, vanishing into upstate New York after his marriage had broken up.

Upstate New York is not that far from Manhattan, but Patty and I rarely saw our uncle. There were no letters or phone calls. I could recollect only a single visit, when he was in hospital. He attended neither his mother's nor his father's funerals, and my parents' conversations about him were conducted in muffled tones. I knew that occasionally he touched my father for fifty dollars, and that he had collected his share of his father's modest estate, but he scarcely entered our lives. My mother never mentioned him. Few people knew that she had a brother. 'Never discuss your business with anyone,' she would say to us. 'Everything is private.' And no subject was more private than her brother.

But Pierre, and his death, had come into my mind while I was folding away my sister's dresses. For, like Patty's, Pierre's death had been sudden and shocking. He had died by freezing to death in a field. So I asked my mother to tell me more. She gave me his age and the date of his death. He had been a year older than Patty when he died.

I asked about the circumstances.

On a cold night, he had hitched a lift, Mother said, but, proving to be a drunk and disorderly passenger, the driver had set him down. Not realising the danger, Pierre had wandered into a field. At the moment of his death, he had been trying to pull off his clothes, imagining that he was too hot rather than too cold.

I asked a few questions, but Mother would say no more.

I went to sleep thinking about Mother's history. She had been 35 when her mother died, 49 when her father died, and 54 when her brother died. She had outlived them all by decades. Now she had outlived her husband and a daughter. My mother was a survivor.

SUNDAY, 20 SEPTEMBER

During the small hours, I tried to sleep. In my bed, I descended with the divers into that underworld, five miles off the coast of Nova Scotia. Through the void we sank until the lights on our helmets illumed a watery forest of ripped metal, weedy cabling, the drenched possessions of the dead. I drifted along as insentient as if I were dead myself.

At 7 am, my mother was drinking orange juice in the kitchen. 'When I wake up,' she said cheerfully, 'I get up. I wash my face. I clean my teeth. It's always a couple of minutes before I remember.'

Yes, Mother was a survivor.

⚜

In the lift, a woman smiled at me. 'You look so anxious,' she said.

'Do I?' Startled, I considered her. She had long, dark hair. She was pretty, young and fresh-faced.

'It's because I'm afraid of getting stuck in the lift,' I said.

'Oh yes, that's been happening recently, hasn't it?'

At Patty's, I tackled the glass jars packed into the kitchen cupboards and a walk-in closet in her living room. There were small ones, yoghurt jars with plastic lids that snapped on and off. She had given me one once, filled with herbs from Provence. Useful for gifts, but dozens of them? There were big jars too, juice jars. What was she saving these for?

I lugged them down the four flights to the basement, a lonely place in the bowels of the building. Once Patty had been mugged, not in the basement, but in the hall, getting trapped by a teenager with a knife. Instead of relinquishing her purse, she had bargained. In the end she got away with handing the mugger five dollars.

It was not in her nature to let go of anything easily.

Obsessively, she had saved and sorted the freebies that came in her hotel rooms: pens, soaps, shampoos, sewing kits, shower caps. Similarly, she kept the plastic cutlery, straws and paper napkins that come with takeaway meals. She used to arrive in Ireland with sacks of this stuff for me; as soon as she left, I would throw it out.

Her phone rang. It was the lawyer. In passing, he mentioned that Patty's pain and suffering might be considered to be an asset of the estate, and therefore subject to taxation. As he spoke, I slid open a kitchen drawer. Inside were dozens of matchboxes. Swissair matchboxes, neatly stacked. I recoiled, my mind alive with flames. Swissair flames, racing through the ceiling of the doomed MD-11 airplane.

At 8 pm, I left Patty's apartment, carrying the frame of the chest of drawers that had held her underwear. I had decided to squeeze it into the bedroom where I was sleeping, for already I was tired of dressing in the kitchen. I hoped to square it with Mother by filling half the drawers with the condolence letters, which continued to pour in. But as I staggered under the weight, my mind was worrying about something else: keys. Jesse, the antiques dealer, had mentioned that the building's superintendent held a spare set of his keys, so I suggested to my mother that she consider the same.

She was outraged. 'Don't be such a jerk!' she snarled.

'It's not a bad idea. What if you locked yourself out?'

'I won't lock myself out.' Then Mother treated me to a colourful stream of invective that cast doubt on my faculties in every way. I listened until she was done. Then I said, 'Would you not consider it? For my sake?'

'For your sake?' my mother sneered.

I should have known better. Mother rarely took advice, and never from her daughters. In my heart, I had known that she would not give a key to anyone. And it wasn't her locking herself out that worried me. It was the idea of her being trapped in the apartment, and being unable to get help. For I would never forget the morning, only two-and-a-half weeks earlier, knowing Patty to be dead, and imagining Mother was too. I could still hear the ringing phone. That vacant ringing, punctuated by my helplessness.

TUESDAY, 22 SEPTEMBER

In Patty's apartment, an ancient canvas bag revealed watercolours and some sketches. At college, she had majored in Art History, and she had a talent for creating beautiful things. I had urged her to make time for her painting, but she never did, not even when she was off work with her broken ankle.

I couldn't bring myself to throw out the dried up paints, the half-finished sketches. Instead, I hid the bag in Mother's apartment; it was easy to lose in that other labyrinth of possessions.

My sister's apartment looked as if a cyclone had hit it. Boxes and plastic sacks were everywhere. The furniture had been pushed around. Every wardrobe was open, disgorging its contents. Patty would have been furious.

I realised that I would have to work faster, for in spite of the chaos, I had only scraped the surface. The more I did, the more there seemed to do. I was due back in Dublin in early October. All my life I had had recurrent nightmares about not being ready: for an exam, or a flight, or work. In the dreams, precious minutes had ticked away, anxiety rising to a choking point. This clearing out of Patty's apartment, this finishing off of the job that Flight 111 had begun, now seemed to embody those nightmares. How was I going to finish on time?

⁂

My cousin Adam's wife, Emma, suggested that we meet for lunch in a French restaurant, a cheerful place with a lemon yellow front. Over salads and wine, my gorgeous companion talked about architectural preservation. I listened, marvelling. Here I was, having lunch (insofar as my stomach let me) with an interesting woman in a pleasant place. There were none of the dark undercurrents that characterised my mother's company, nor was there an agenda of legal and financial matters to plough through. We were there to enjoy ourselves. The miasma of sorrow that clung to me shifted slightly. I loosened up. I shared my dismay at the things my sister had saved. My companion listened, then explained how she had once cleared out a dead friend's apartment. She offered her help.

I declined. The task was too private, too fraught. However, the offer worked as a psychological boost: I did not feel quite so alone. What's more, Emma proposed a solution to the problem of my mother's keys. 'Why not give a set to Adam?' she said. 'Your mother trusts him.'

It was true. Mother admired Adam, and he lived near enough to act quickly. If there was anyone whom my mother might trust with her keys, it was Adam.

I hurried back to the coalface of Patty's apartment. I carted books down to the steaming basement; I jammed papers through the shredder; I hauled sports gear out from the bottom of a wardrobe: Patty's brand new ice skates; a trio of tennis rackets. On one racket there was a familiar, needlepoint cover. I had watched her sewing that cover, her long fingers pulling the threads through. The cover read: LOVE. A tennis word, of course, but also something more complicated.

Patty had lived her life yearning for love, in particular the love of a flawless man. She dreamed of a man who had been to the right schools, who wore the right clothes, who worked in the right place and made plenty of money. Two comfortable homes (city and country) came with this paragon, and two perfect children. Compromise did not interest Patty. She believed in a perfection that did not exist.

She had wanted me, her sister, to love her too, but I had failed her like everything else. I had not been the person that she wanted me to be. I could not even provide the companionship that she wanted, because I didn't measure up to a standard that was alien to me, to which I couldn't even aspire.

LOVE, I thought, looking at the racket cover that she had created and left behind. By chance, 'Love' was the name of the short story I was writing on the day that she died. The story I had left unfinished on my desk.

I raced back to Mother's building, lugging two of Patty's wire chickens, filled to the brim with seashells. They were for the lawyer, a keepsake of Patty. My mother and I met with him in the lobby, because of her hostility to visitors in the apartment. There, under the scrutiny of the doormen and passers-by, we conducted our private business. The lawyer had a sheaf of pow-ers-of-attorney: my mother for my father; my sister for my father; my sister for my mother; my sister for me. One by one, he tore them up, death having made them redundant. He had prepared a single new one: myself for my mother. Without hes-itation, she signed. 'It's terrible to give one person so much power over another,' she said.

When the lawyer got up, I handed him Patty's chickens. They pleased him, he said, chickens made from chicken wire. And so he vanished in the heat, doubly burdened now, with his heavy case and the bagful of chickens.

By the time we went upstairs, I was exhausted. I climbed hastily into the bath. We were going to the opera and I wanted to be clean. For thirty years my mother had subscribed to the opera. Only five months earlier, we had been to *Carmen*, and since then, I had been preparing for Handel's *Partenope*, by listen-ing to it on CD. I had never thought about going to the opera without Patty. I had assumed that we would have pizza and root beer beforehand, as we always did. I had imagined that these pat-terns would go on forever.

My mother was wearing a polo shirt with a Mickey Mouse insignia, not conventional opera dress. 'It was the last thing I bought with Patty,' she explained. 'On our last shopping trip.'

We were on the cross-town bus, heading for Lincoln Center. With gusto, my mother described the fights she had been picking with people in the street that day. 'The first was a Jewish girl with her skirt way up around her behind. It's Rosh Hashanah. Isn't she supposed to be thinking spiritual thoughts? "Your skirt's too short!" I told her. Then later I yelled at a woman on a mobile phone. "Those stupid things!" I said. "You're invading my space!"'

When we got off the bus, we were surprised to find the antiques dealer behind us. My mother had invited him to use

Patty's ticket. Jesse was smartly turned out in an ivory double-breasted suit, with bowler hat and umbrella. We were early, so we sat in the park, where Jesse fell in easily with my mother's complaints about the masses. His own *bête noir* was taxi-drivers. 'The other day,' he said, 'I walked right through a cab blocking Second Avenue. I opened one door, climbed out the other, and left them both open.'

In the restaurant, Jesse tucked into his juicy slab of beef. I had known him to see since I was a child, but I knew almost nothing about him. A few questions proved rewarding. His personal history was lively, textured and full of surprises.

Around us, a sea of busy tables, laughter and intimate conversation. I listened and watched, then drifted loose, until I saw myself, a middle-aged woman dressed in black, struggling to eat, a visitor in the ordinary world.

Handel's opera began, the gorgeousness of the set undermined by ugly costumes. The title character was squeezed, sausage-like, into an alarming pink knitted suit. Soon, nothing mattered except the music, in particular the extraordinary sound of two counter-tenors, a powerful, unearthly pulse that engulfed me for a few, welcome minutes.

But the second half of the opera proved to be too much. All I could think about was Patty. She was meant to be sitting next to me, complaining about jet lag, planning our next weekend in Connecticut, infusing the evening with her own inimitable flavour. Unseeing, unhearing, I sat in the dark, blotting at my streaming eyes and nose.

Jesse removed his bowler hat as we boarded the bus that would return us to the East Side. 'I don't like people looking at the back of my neck,' he said, 'because the hairs on my nape don't grow in an orderly fashion.'

We lurched into Central Park, past the twinkling lights on

the trees outside Tavern on the Green. The bus groaned and whined. My mother, I saw, was riding the ghost bus of the past. We lumbered onwards, across Fifth Avenue, down 64th Street. Then the bus pulled left onto Madison. It stopped at the spot where Patty would say goodnight and slip off. My mother looked at me. We shared our unspoken thoughts, as we continued homewards.

WEDNESDAY, 23 SEPTEMBER

Deep in sleep. Terrifyingly deep. Thrashing and gasping, I shuddered into wakefulness. I was still alive.

It was 3 am. I decided against going back to sleep.

7 am. Mother was making her instant oatmeal. Her eyes swept over me critically. 'Everybody else combs their hair,' she said. 'Why don't you comb yours?'

On my way to Patty's apartment, I stopped at a drugstore. My mother had been hunting for a favoured hair tonic. I poked around the shelves. Two dusty bottles emerged. I bought them both, and some Starbuck's coffee. My eye drifted towards the yoghurt. It was Swiss-style. I recoiled. Everything Swiss filled me with horror.

In Patty's apartment, two freshly ironed shirts were still hanging from the wicker surround above her bed. They shifted slightly in the air currents. Phantom shirts. Had Patty known that she was about to die, surely she wouldn't have ironed shirts. What would she have done instead?

Extraordinary things tumbled from the crevices of her apartment. A satin handbag that opened to play an aria from *Faust*. Black, shoulder-length gloves and a fusty old muff. Fur boas with heads that bit one another's tails. My mother's balding, sealskin coat, stinking of mothballs. As a child, I had snuggled up against it, the fur soft on my cheek. I manhandled these treasures, these ones-of-a-kind, stuffed them into sacks, ravaged my sister's belongings like some modern Visigoth.

A bundle fell into my hands. I opened it. Letters, guidebooks, notes spilled out: everything about Iran. Iran? Suddenly I remembered that Patty had lived there for a couple of months, flying pilgrims to Mecca. How could I have forgotten? And was I now going to throw out her precious memories? I hesitated, then shoved the bundle into the ballooning bags of personal papers to be saved and shipped to Ireland. When would I get the time to go through them? But who would remember all the important things about Patty if I didn't?

I worked on frantically, pillaging and desecrating. Ivy the Hun. Hours passed. My mother phoned once. Then a second time. 'When are you coming back?' she asked. 'I can't wait around all night. I've got to eat.'

'Go without me,' I said. 'Just tell me where you're going.'

'Kentucky Fried Chicken.'

My stomach lurched.

'Don't stay in that apartment after eight o'clock,' Mother said. 'Promise.'

I promised. What would happen if I stayed? Would I turn into a pumpkin?

Eventually, I locked up, my arms heavy with offerings for the garbage cans of New York, which I distributed munificently.

In the lobby of Mother's building, a couple of letters from my Irish friends were waiting. The apartment upstairs was empty. How lovely it would be to sit alone, reading my letters and sipping a glass of wine. Instead, I headed out again, drawn like a moth to the candle of Kentucky Fried Chicken. Through the plate glass window, I could see Mother, alone at a melamine table. In front of her, a cardboard box gaped. The plastic lid of her styrofoam cup was pierced with a straw. I watched her shaky hand lift the plastic fork. Her false teeth chewed. Her eyes, filmy with cataracts, protruded from the leathery sag of her skin.

Winter was coming. The sidewalks of Manhattan would be cold and inhospitable, bitter for a bereaved old woman.

The lights in Kentucky Fried Chicken were yellow. Garish signs, touting buckets of chicken, plastered the walls. Behind the counter, the young workers hid neither their boredom nor their contempt. What was my mother doing in there? Why did she refuse to dine more graciously? At least Patty would never sit alone at the age of eighty, eating plastic food in a fast-food emporium.

I went in and sat down. 'Hello, Mrs. Eberhart,' I said.

Mother's face creaked into a smile.

'Look,' I said. I placed the two bottles of the favoured hair tonic on the table in front of her.

'You found it,' she said. 'Wonderful.'

I took out my letters from Ireland and read them to her, as she finished her dinner. Then I described some of the treasures I had found in Patty's apartment. We walked home in a loop, ambling slowly in the dark, looking in every window, having as good a time as possible under the circumstances.

'Patricia asked me to come to her apartment,' Mother said. 'She wanted me to help her throw things out. I said, "No! Do it yourself!"'

An impossible scenario, I thought. The one hoarder advising the other on what to throw out. It never could have happened. Not if they both lived a hundred years.

THURSDAY, 24 SEPTEMBER

7:30 am. I was kneeling on the floor, busy with paperwork, using the bed as a makeshift desk. My mother came in. In her hand was a white circlet with flowers twined around it, a bride's headdress. A tag fluttered with the message: Lasting Promises. 'I found it in Wal-Mart,' Mother said, 'the last time we went shopping. I showed it to Patricia. "Want to get it?" I asked, "so you'll be ready?"

'"Oh, Mother!" Patricia said. She shook her head. So when she wasn't looking, I bought it. Just in case. It was only three dollars.'

Just in case. Patty had spent her lifetime getting ready for the wedding day that never happened. There was more than enough china and crystal in her apartment for any bride. The china was Rosenthal, the crystal Waterford. The blue and white cups were kept on open shelves, but when I reached for them, they would not budge. Time and heat had welded them fast to the paint. With difficulty I eased them free, leaving behind the odd flake of china. These cups were a fraction of Patty's trousseau. In the closets and cupboards, a dinner service for twelve revealed itself, with 36 matching crystal glasses, each one wrapped in tissue. There were dozens of further pieces: a Limoges tea-set; antique plates; half-a-dozen blue and white teapots.

By contrast with this abundance were the things in the dish drainer, which Patty actually used: a pair of chipped French coffee bowls and a recycled paper cup. It made no sense to me. Why collect beautiful things if not to use them?

A pile of papers was balanced on a three-legged table. On top was a fat volume, *Guide to Careers in World Affairs*. Beneath it was a folder of articles, all annotated in Mother's hand. Names of companies and people were underlined, sometimes twice or three times. 'You could work here' was scribbled in the margins. Or, 'Why not a career in banking?'

At the time of her death, Patty had been flying for nearly thirty years. After college, she had worked for Pan Am. When Pan Am went bankrupt, she went to Delta, although she could have chosen a severance deal instead. She never stopped talking about leaving, but she had no self-confidence. Yet she would have succeeded at many things. Attentive to detail, conscientious and hard-working, she would have been an asset to any employer.

However it was more than fragile self-confidence that kept Patty in the air. She had difficulties with changing. I believe she would have stayed in whatever job she had taken after college. Similarly, she never moved from her building, although she talked about it constantly. She couldn't let go. Not of a glass yoghurt jar, or of shoes she never wore, or of the job at which she had worked for thirty years. At some fundamental level, she was unable to abandon the devil she knew.

That would have been the hardest thing for her in that airplane, I thought: having the greatest change of all thrust upon her.

When I walked into my mother's lobby, the doorman handed me a Fed-Ex. Inside was a temporary death certificate, the second I had received. Like the first, it was incorrect. Patty's place of birth was entered as Brooklyn instead of Norwalk, Connecticut. I sat, for a while, studying the document. The very idea of a *temporary* death certificate was ludicrous. Was it expected that the dead would come back to life?

The certificate stated that Patty's remains had been recovered on 3 September, the night I had watched the news in Dublin, with its clip of a body bag being unloaded at a wharf.

Had it indeed been Patty? I thought of her torso, presumably naked, with its part of a head, and its one ragged thigh stump, longer than the other. What had those emergency volunteers felt, as they hoisted it over the side of their boat?

The cause of death was listed on the certificate as *Blunt Trauma*.

※

For dinner, my mother chose a Chinese fast-food restaurant. Behind the counter, two men laboured sweatily over steaming woks and cauldrons of rice. A neon light twitched overhead. I sat at a greasy table in the heat, picking at a heap of shrimp, vegetables and rice, enough to feed a small family. Outside the open windows, traffic thundered past.

My mother's taste in restaurants was getting me down. We could be enjoying the airy quiet of the roof garden on top of her building, eating fresh and wholesome salads. We could be in the apartment itself, taking a simple meal at a magically cleared table. And Manhattan was full of nice places to eat. But no. This is what my mother insisted upon: lining up for fried food as if we were in boarding school or prison, eating with plastic cutlery, drinking out of cans.

As always, she talked more than she ate. Then, without warning, her monologue took a surprising twist. 'For ten years,' she said, 'after Patricia came back from Florida, she kept saying that I had abused her.'

A fire engine tore past the window, its sirens screaming. Had I heard her correctly?

'What did she mean?' Mother said. 'What did she mean that I had abused her?'

My hands gripped the table. I knew exactly what Patty had meant. Not abuse in its most ghastly, physical manifestations, but something more subtle. Something that had stained and twisted the fabric of our whole lives: the way Mother treated us. It was her world of verbal violence, of unstoppable invective. Her relentless assault on our behaviour, our looks, our tastes, our intelligence, our abilities, on anything and everything that we were or weren't. As a child, I retreated, hid. Patty, as the elder, stood her ground, bore the brunt, fought back. By the time she was a teenager, she could – and did – yell just as loudly as Mother. While growing up, I took this tumultuous underworld for normality. As a young adult, experience suggested other ways of life, and propelled me into putting as big a distance between myself and Mother as possible. Dublin was barely far enough. 'You drove her away,' my mother complained openly to my father. 'It was your fault! Your fault! You were such a peasant!'

Mother was still looking at me. The neon light twitched like a strobe on her perplexed face. What did Patty mean, my mother had asked. I knew and my gut clawed at me.

Ten years, Mother had said. So why, during all that time, had Patty not said the same thing to me? Was it that I lived in Ireland? But we saw each other regularly and wrote letters. Surely something might have been said over ten years' time? A long opportunity to have an adult conversation, the kind we never had, not once in our lives. Had Patty said it to me, we might have staggered towards understanding, groped towards accepting one another as equals and friends.

'What did she mean, that I had abused her?' my mother said yet again. What could I answer? What could I say to this tragic old woman? Under these appalling circumstances? In this awful restaurant? That it was too late? That what had been destroyed could have been saved? That our lives should have been different? That Patty and I might have become adults together, and enjoyed better lives through each other? That two sisters should have been the best of friends?

I looked at my mother's sad face.

I said nothing.

⚜

I left my mother at her door, then went to hunt out the *The New York Times*. When I returned, I heard her snarling into the telephone. 'Ivy,' she said rudely, 'you don't know anyone by the name of Alan Crawford, do you? There's this person who says that he knows you.'

'Of course I do.' I took the phone. 'Hello, Alan,' I said warmly. But the caller was gone, and I was speaking to air.

'Oh, Hortense, you are *so* nasty,' my mother said to herself with some satisfaction.

Alone in the bedroom, I picked up the Manhattan telephone directory. There were numerous Crawfords, with various possibilities. And I had no reason to believe that my old friend had even been calling from Manhattan. I closed the directory. There was no point. If I did manage to speak with him, my mother was bound to get angry.

⚜

10 pm. My mother and I were in the living room, peaceably reading. 'Aren't you glad that you didn't go to Nova Scotia?' my mother asked.

'No, I wanted to go.' I had wanted to see where Patty died, to talk with people, to understand at first hand something of the terrible process of recovery of the dead, their possessions and the wreckage of the airplane. However, even as the few honest words slipped out of my mouth, I knew that I had made a mistake. My mother flew into an instant, ungovernable rage. 'Shit!' she yelled. 'Shit! You can go to hell! You're a jerk!'

Yes, Mother, I should have said. *You were absolutely right, Mother. I am glad I didn't go to Nova Scotia.* For she had not asked her question to elicit my feelings. What she had really wanted was to have her own action (or inaction) reinforced.

And I had blown it.

She raved at me. 'You're full of shit! I used to have you under my thumb, and I loved you so much. Now you're just a jerk like everyone else. Horse's ass! Shithead!' Profoundly upset, I wandered off to my bedroom and began scribbling down her words, a way of detaching myself from their impact. There was no problem hearing her. Her emotional tempest filled the apartment, shook the windows. Anything I said would only make matters worse.

'This is just such a waste of my life,' she bellowed; 'one pile of *scheisse* after another.' Papers began to rustle as her invective rattled on. 'They're alive,' she snarled. 'They're full of fire. They're pressing us off the edge of the earth.'

I realised that she was looking at doctors' bills. 'They charge you an arm and a leg and a foot. And if you dare to challenge them.... Hah! To be treated like I'm over the hill. I want them to know that I'm a real pain in the ass.'

The clergy came next. 'So the minister calls on me. Jesus! What can she do for me? She comes to offer me succour, then she sends me the bill. Well I'm not going to let them push me around. They'll hate me by the time I'm finished with them.'

On and on she went. Eventually I heard her howl, 'I want my best friend back. Oh, Patricia, Patricia! Why couldn't you have stopped being so nuts? Why couldn't you have gotten out of that stupid business?'

After that, she was silent.

In the morning Mother wandered in while I was making my bed. She made no reference to the night before. It was as if she had forgotten her terrible fury that had kept me awake all night, my stomach in an uproar. She took a black balloon of a dress off the crowded exterior of a wardrobe. 'Try this on,' she said. 'I ordered it by mail order. It's too big for me.'

The dress flapped, catching ingloriously on my derrière. I shook my head. 'Not me,' I said politely.

'Stunning,' my mother insisted. 'Take it.'

We had been to see the first of what I called the sharks, the legal firms that specialise in plane crashes. In a huge, minimalist office, we had listened as a guru of aviation disaster litigation described his own expertise. Flanked by a posse of what he called his 'team', we had sat at a circular table. At the touch of a button, its heavy glass surface moved up and down.

Now we had two more appointments, scheduled back to back. The first proved to be a pleasant surprise. A father and son spoke quietly with us in their library. Unassuming and sensitive, they related to Mother as an individual: they connected with what Patty's loss meant to her. What were they doing in a business like this?

The next lawyer, a breakaway from the guru, was tough, long and lean. He himself had been a pilot in Vietnam. 'May I call you Hortense?' he said to my mother, but, at once picking up her reserve, he seamlessly addressed her as Mrs. Eberhart. Later, she made a little noise. 'Is that a cough or a sneeze?' he demanded. It was so bizarre and surprising an interjection, such a splitting of hairs, that I almost laughed out loud. Energetic, sharp-minded and a little bit scary, this man was cut out for his job.

Afterwards, in the lift, it was obvious that Mother, myself

and our own lawyer were in agreement. The father and son were our kind of people. I voiced one reservation: that if we were being hard-headed, perhaps the former pilot would be a better choice because he was tougher. The lawyer reassured me. In practise, he explained, the outcome would be similar; we should choose the people we wanted to work with.

We stood on the street finishing our discussion. My mother turned to the lawyer, 'Doesn't my daughter wear terrible shoes?' she said.

Startled by this shift in the conversation, the lawyer glanced at my feet. I was wearing laced walking shoes, round-toed and comfortable. 'They look alright to me,' he said.

I fled uptown at a pace. Already it was after 1 pm.

⁂

I tackled Patty's books and papers. There were a hundred French paperbacks, which Patty had read to maintain her fluency. Each page was heavily annotated with English translations. She had read every word for absolute understanding, transforming each book into a palimpsest that reflected hours of reflection. I threw the lot out, except for Marguerite Yourcenar's *L'œuvre au noir*. Annotated to page 311, it was the one that Patty had been reading and would never finish. I saved her biggest French dictionary for Richard, and her Jansen – an art history textbook – for myself, but disposed of other language and art books cavalierly. Like a windstorm, I rifled her travel books, each fat with notes and maps, records of her extensive travels, bundling the lot into sacks.

My rout through Patty's non-financial papers threw up surprises. A dozen notebooks crammed with Italian. Notes from the New York Power Squadron, where she had studied Boating, Sailing and Navigation. Folders full of Financial Accounting course work. And the relics of countless other courses, exhaustive evidence of her will to self-improvement. Patty had saved everything. I chucked it all out. Even worse, I shredded almost nothing. The quantities were overwhelming, and time was running out. She would have been livid. I was furious myself, for desecrating the lifetime of her knowledge in a few, frantic hours.

I worked harder and faster, whipping through hundreds of books and folders. I dragged heavy bags of paper and print down the hot stairs to the hotter basement. Finally, at 8 pm, I readied myself to go. A friend of Patty's lived on my way home. I wanted to give her a pair of watercolours, so I phoned to say that I would leave them with her doorman. In the middle of a sentence, I broke down. The woman kept calm. 'Go ahead,' she said. 'Cry all you like. It's good for you.'

I locked up. The paintings were heavy. I struggled downstairs. On the landing were two old women: brightly dressed, heavily made up and stinking of drink. Patty's neighbours. Old bats, I thought, harpies.

They had been old back in the days when we were kids, when my father had managed the building. My sister loathed them, a compliment which they had returned. Stunned, no doubt, by having outlived her, they cackled their regrets at me. I was not impressed. One of them had been quoted in the newspapers, complaining about the disturbance of journalists calling to the door.

Their beady eyes followed me downstairs.

Vultures.

With the paintings, I walked into the unfamiliar building. 'You're Patty's sister,' the doorman said, instantly recognising Patty's features in my own. He began to sing her praises. People adored Patty. She remembered them, took an interest. Had this doorman, I wondered, been given some of her home-baked cookies? At any rate, he was in a talkative mood. He moved briskly away from the horrors of plane crashes to the horrors of what he called Nam, which he had experienced at first hand: two years and two days of it.

Half-mesmerised, I listened.

'It wasn't the bugs or the heat or the death. No. None of that was the worst. It was that there was no purpose. It was for nothing. There were one hundred and fifty men in our unit. Only forty came back. And what did we do? We took, then lost, then took the same little hill, over and over again.'

I was fascinated, but so tired that I could scarcely stand. With difficulty, I detached myself, and hurried on home.

SUNDAY, 27 SEPTEMBER

On my knees at my bed-desk, I was opening Patty's mail. In it was a summons in connection with some property. I put it aside with a view to asking the lawyer about it at the next opportunity.

'Call him now,' Mother said.

'It's Sunday morning.'

'Call him now.'

'I can't call him at home on a Sunday morning. You don't disturb people at 9 am on Sunday morning.'

'I'm telling you to call him now.' Mother's voice was ascending the scale dangerously. As far as she was concerned, the lawyer was to be telephoned immediately. Not Monday. Not in a couple of hours, but now.

I dialled. After half-a-dozen rings, a woman answered sleepily. I apologised. 'It was time to get up anyway,' she said nicely. And for all that, the lawyer wasn't there. He was out playing tennis.

My sister's linen closet was full of things that had never been used,

towels and sheets that were still in their wrappings. In her bathroom, the towels that she did use were ragged. The sheets on her
bed were threadbare. A gorgeous shower cap popped out of a storage basket. I had bought it, twenty-five years earlier, at Liberty's in
London. A student at the time, I had agonised over its cost. Yet I
had wanted to give Patty something special, something beautiful
that she would use every day. And here it was, untouched, hus-
banded, like everything else, for a future that never came.

In a drawer, three delicate nightgowns were folded into tissue. On the back of her bathroom door was the nightgown she
wore, mint green polyester, with the stitching coming away.

There was no stop to the heartbreak of unravelling Patty's possessions. A dozen folders were filled with meticulous paper cuts:
birds, wreaths, flowers, angels, every wispy detail faithfully achieved.
Each piece was a work of art, the fruit of half an hour's labours.
How could I throw them out? But what would I do with them?

A basket full of gift wrappings reflected her passion for giving,
but other collections were harder to fathom. Tickets, for example,
popped up in their thousands. Scraps of paper from the Paris
Metro, the London Tube, the Moscow Underground, Dublin Bus,
the Réseau de Nice. Hundreds of New York City bus transfers.
Tickets from museums, galleries, châteaux, botanical gardens,
from all over the world. No doubt, I could have tracked her life
through those tickets. Instead I threw them out, together with the
cigar tubes from a kitchen drawer. One was full of little springs:
from the insides of cheap, ballpoint pens. The other held the metal
rings that join the two halves of the pen. Waste not, want not.

I closed my eyes, reached into the jar full of Chinese fortune
cookie fortunes, and picked one out. 'Don't let unexpected situa-
tions throw you,' it read. I tipped the rest of them into the rubbish.

My still disturbed digestive system had had enough. If I had
to eat out, I wanted to be waited on, in a pleasant ambience that

wasn't lit by fluorescent lights. I wanted to be surrounded by families and couples, not gangs of unruly teenagers.

'Can we go someplace else tonight?' I asked Mother.

'Where would you like to go?'

'Patsy's,' I said, referring to an Italian restaurant where I had eaten with Frank. The food was good, and the place gave off a seductive waft of basil.

'Why would you want to go there?' Mother asked suspiciously.

From the moment we sat down, she made trouble. She peered at the menu. 'Too expensive,' she growled. When the waitress arrived, Mother grilled her about every entrée, an inquisition connected in theory with her heart medications, but Mother never asked questions in Kentucky Fried Chicken. 'I can't eat a thing on this menu,' she complained, but eventually she ordered. When a ripple of laughter broke out at a nearby table, she snarled, 'Too noisy!' And when the food arrived, she consumed it with a dramatic reluctance that suggested she was ingesting fish eyes.

We did not have fun.

I tipped the waitress generously. It wasn't her fault that we wouldn't be coming to Patsy's again.

WEDNESDAY, 30 SEPTEMBER

Mother was poring over Patty's bank statement. 'I don't understand this,' she said. Her fury with the bank could not be contained. 'Jackasses!' she yelled. 'They think they can do what they like.'

'Ssh,' I said, concentrating on the figures.

'Don't patronise me,' she roared. 'You don't treat me right.'

'I'm not patronising you,' I said. 'I'm just asking you to shut up so I can work this out.'

***

The movers arrived at Patty's to pack up the mountain of belongings that was going to Ireland. Working steadily, the boss revealed himself as a man with a philosophy. For the last few years, Bill said, he had been down-sizing. He had sold his house and car. He kept no bank account. He owned nothing, except what he used immediately. 'I live my life,' Bill said. 'I take it like it comes. I go to work; I go home and chill out. I don't clutter up my life with stuff I don't need.'

I listened wryly, as Bill got on with it, up to his armpits in Patty's possessions.

***

The lawyer tossed an envelope into Patty's wastepaper basket, only to watch me retrieve it and feed it through the shredder. 'Does your mother still shred everything?' he asked.

I nodded. I told him how she distributed the shreddings, a handful at time, among the garbage cans of New York.

The lawyer shook his head. He packed the rest of Patty's financial papers into a trunk. Then he squeezed himself into the tiny lift, with the heavy trunk clutched to his chest. By the time the lift got to the ground floor, I was already there. Through the glass window, I could see the lawyer's profile. Downstairs, the lift door opened on a different side. I hauled it open. He turned, astonished, whether by me, or by the unexpected door, I wasn't sure.

***

It was ugly from the moment I came home. Mother was staring at her papers. 'I forbid you,' she said, 'I absolutely forbid you to give my keys to Adam.'

My heart sank.

'Leave me alone,' she snarled. 'Can't you see that I'm paying my fucking doctors' bills?'

I wanted Adam to have those keys. There had to be someone who could gain access to my mother quickly. 'Please, Mom,' I pleaded. 'For my peace of mind?'

'You can go to hell!' she snorted. 'I don't care about your problems. I've got to look after myself.'

'What if you fell?' I said bluntly.

'I'd drag myself to the door.'

'How would you open the door?'

'I'd open it! I'd just open it!'

And so it went, my mother's rage growing hotter, until I gave up. Reason was futile. My mother had always done – and would always do – things her own way.

THURSDAY, 1 OCTOBER

Now there were three movers in Patty's apartment. They were wrapping the large pieces of furniture, and lugging them down the stairs. Mother had said that under no circumstances was I to clean the apartment. So I stood there ankle deep in dust balls, in the wreckage of what had been my sister's home.

My mind moved, as it had every day, to the terrors of Patty's death. Yet again, I considered the framework of events from the first whiff of smoke to annihilation. One question kept nudging others out of the foreground. At what point, exactly, did Patty realise she was going to die? Smoke in the cockpit? With the aircraft at cruising altitude and its fuel tanks brimming? Patty was bright, and she knew airplanes, and she was pessimistic. It seemed to me that she must have known from the beginning.

Still, I thought, there may have been an interval of hope, perhaps when retrieving the dinner service. Possibly then, as she struggled with trolleys, she believed in a Halifax landing for a moment or two. But within seconds, her mind must have run rampant.

In her apartment I had found all her flight manuals, and every update from regular emergency training sessions. I had discovered neat diagrams of airplanes floating on the ocean surface with slides and rafts, and life-jacketed survivors tidily lined up in dinghies. Comical pictures. Absurd pictures. I knew that Patty didn't believe in them. More than once, I had heard her talking about crashes into the Atlantic. The Atlantic Ocean was no runway, and planes did not land intact upon it.

Once she had told me about landing in Tokyo during a monsoon. She was terrified, she said. She had pleaded in her mind with the captain. Just put this plane down. Anywhere. On a road, in a rice paddy, anywhere. But get it out of the air.

I had experienced my own terrors. I had dreaded flights for months before taking them. I had woken in sweats of fear. This fear had taken me over to such an extent that I had gone to a hypnotherapist for help. But as I waited in the remains of Patty's apartment, I realised how my fear was only a pale imitation, only a scrap, of what she had endured on that dreadful night. Patty and the 228 other souls on board.

I don't believe this, I heard her saying. *I don't believe this is happening.* Not on Swissair. Not on the airline she saw as perfect in every way.

My mind dwelled on her fear. A monolithic fear that consumed everything, that stopped the heart, froze the breath. I felt the giant aircraft juddering in the dark. I heard the sick whine of its remaining engines. I watched the final plummet into the sea, and saw the great pillars of boiling seawater, reaching up towards the black, black sky.

❧

There was a small balcony at the back of Patty's apartment, accessible through the bathroom window. For a while, I looked out the window. The view was not unpleasant. There were trees and gardens, and the backs of the buildings on the next street. The leaves were beginning to turn: to yellow and orange and brown. As I watched, a gust of air lifted a branch which rattled against the window.

❧

One of the movers told me about his heart problem. Even at rest, his panting was audible. His generous tummy sagged over his belt; his forehead glistened with sweat. Perhaps, I thought, lugging furniture up and down stairs was not an ideal occupation for him.

What a struggle life was.

The mover had a story. His sister was living in Ireland, married to an Irishman. When they came to visit New Jersey, the Irishman fell in love with the American refrigerator. So he bought one, a top-of-the-range model with ice-maker and water cooler. For a small fortune, he shipped it back to Ireland, where he discovered it was unusable on account of electrical incompatibility. So he had dropped it into the sea.

❧

Late afternoon: Mother's apartment. From the window, I watched the movers unload furniture onto the footpath: Patty's wicker bed-head, her armchair, the quaint little desk we had used as children. The sight of these familiar things thirteen storeys below was surreal.

That night I sat at the desk, returned again to the bedroom that Patty and I had shared for the first seventeen years of my life.

The desk was tiny, but I could sit and write there, or even eat. By contrast with the rest of the messy apartment, it was good. Vigorously, I told Mother not to put anything into the desk or onto the chair, that it was a working desk. She watched, as I arranged a few photographs on its upper ledge.

'Those pictures are going to collect filth,' she said with a lemony look.

FRIDAY, 2 OCTOBER

A month after Patty's death, and I was returning to Ireland. My feelings about going home were mixed. I couldn't wait to see my husband and kids. I wanted to sleep in my own bed and to eat my own food, both of which, I hoped, would cure my battered insides. At the same time, I was reluctant to leave because of Mother. She had lost the daughter who was her only significant social contact. When she wanted to, Mother could be perfectly charming, but she was highly critical of anyone and everyone. The give-and-take of friendships, the sharing of problems, were not for her. Without my sister, she would see nobody.

It followed that the converse would be true too: that nobody would see Mother. Nobody would register her comings and goings. I didn't doubt that she could disappear for a long time before anybody would notice. And she was eighty years old.

On my last morning in the city, I went to Patty's apartment. With a large black marker, I left messages for Mother, advising her what to do with the few remaining belongings. Then I stood in the empty, echoing space and wept. Was it true? Was she really dead? Since Patty's death, I had scarcely stopped moving. What time had I had to reflect?

I prepared to leave for the last time. I gathered my bags and thought of Patty doing the same, a month earlier, organising her flight bag and checking her handbag. With Patty, I switched off the air conditioner and the lights. I felt her hand on the doorknob as I turned the locks. Together we descended the stairs. I thought of her innocence, of her never suspecting that this was it, that in a few brief hours she would be a ghost. Unlike Patty, I had the melancholy luxury of knowing that I would never return.

I opened the main door, passed the mailbox with her name on it, stepped out on to the street. A month earlier, she had turned left for the last time, to walk the few feet up to Madison, where her driver had collected her and taken her to the airport. I hovered on the doorstep for a moment. Then I turned right, towards Park Avenue.

⁂

I left for the airport with hours to spare. Most seasoned travellers, I understand, choose to leave at the last minute, but I prefer to approach the skies in a leisurely way. I was no longer afraid of flying. What did I have to be afraid of, compared to those who had died? Perhaps, on some subconscious level, I had known my life would be changed by an air disaster. Now that it had happened, the fear was gone.

I also had a profound need to be alone and doing nothing, with nothing expected of me. When I got back to Ireland, there would be things to catch up with, and the expectations of my family, especially my younger son. This flight would be my breathing space, my time-out to think quietly.

The woman from Delta came and found me in the Aer Lingus lounge. She handed me some forms about the disposal of Patty's remains. We chatted politely for a while. Gradually I realised that the woman intended to wait with me until my flight. Tactfully, I thanked her for taking the trouble to come all the way to the Aer Lingus lounge. Surely she had plenty to do, I said nicely, and I was going to be OK.

She left.

I waited in silence, drained of energy.

The airplane took off. The lights of the greater New York area twinkled below, as magical a sight as ever they were, no doubt as magical as they had been on Patty's last night. Had she looked at them? Had she marvelled at them, let them fill her with wonder? About an hour into the flight, I did not need the captain's announcement to tell me that we were passing over the coast of Nova Scotia. I felt the location throughout my body, a shuddering grief.

Entracte

It was 6:30 am local time when I disembarked at Dublin Airport. Incredibly, the man from Delta who had accompanied me to New York was waiting at the door of the aircraft. He drove me across the city, deserted at that hour on a Saturday. I was relieved to be home.

⚜

The telephone rang. Would I do an interview on radio, in connection with a public lecture that I would soon be giving? The disembodied voice and the question seemed bizarre. I had just gotten off a plane, I had spent the last month immersed in death, and here was a man, blissfully ignorant, asking me to talk about my research.

I hesitated. 'Will it be live or recorded?'

'Recorded.'

'Yes,' I said. 'I'll do it.'

A few days later, I was in studio. The interview went smoothly. Afterwards I headed for a coffee morning. I parked a few doors down, but stayed in the car. To go to a coffee morning, while Patty's remains were on ice in Canada, hardly seemed decent. After a while, I went in, made small talk, listened to some woman's account of her messy separation. I left as soon as it seemed polite.

Later, when my older son Richard and I were exchanging notes about the day, I explained how I was trying to do normal things. 'But, Mum,' he pointed out, 'you wouldn't normally go to a coffee morning.'

In fits and starts, I tried to re-enter my life. In Samuel

Beckett's play *Waiting for Godot*, Vladimir asks, 'What do you do when you fall far from help?' Pozzo replies, 'We wait till we can get up. Then we go on.'

I went out to dinner with friends, a dinner that had been postponed on my account. As I sat there, it struck me that I had become two people. I was still myself, the person who observed intently and occasionally jumped into the throes. At the same time, I was an outsider from a world of shadows, drifting along the edge.

Mother telephoned from her house in Connecticut. Since Patty could no longer drive her, she had walked to Grand Central Station, taken two trains, then walked a couple of miles more, carrying her suitcase, milk and orange juice.

In the washing machine, she found Patty's gardening clothes, waiting to be washed.

⚜

My tooth began to throb. It hammered and thumped. I almost enjoyed the pain, so thoroughly did it occupy me. But after it was attended to, the emptiness came back.

On All Souls' Day I went to church, St. Bartholomew's in Ballsbridge. Everywhere inside, I saw Patty: in the front pew, as I came up the aisle in my wedding dress; at the font, holding Richard to be baptised; beside me, singing carols on Christmas Eve.

⚜

Richard handed me a Swissair press release, downloaded from the Internet. It said that Swissair had disconnected the in-flight entertainment systems in their fleet of MD-11s as a 'precautionary' measure. The personal consoles, installed at great expense, had touch-sensitive screens, and they allowed passengers to gamble with credit cards.

For the first time since the crash, I became angry. It was one thing to believe that the disaster had been caused by some failure in systems that were necessary to fly the plane. It was another to think that 229 people might have died for in-flight entertainment.

I could not apply myself. I sat at my desk and hammered out a few sentences, then my concentration wandered. My failure to get things done demoralised me. I abandoned all pretence of working. Instead, I painted the living and dining rooms. The freshly painted walls, at least, justified my existence.

10 November. My mother telephoned. The sink in the kitchen in New York was blocked, she said, but that wasn't why she had called. The real reason was that a cousin had said that her grand-daughter was in Ireland, and that it would be nice if we could get together. My mother didn't like this. 'I hope that you won't see her,' she said. I tried to say that I liked meeting people, but Mother would hear none of it. 'If you had called her while you were in New York, they would have considered it extremely gauche.'

When I related this conversation to my husband, he shrugged. 'Your mother has chosen to be alone, and she wants you to be alone too.'

The young woman telephoned the following morning, and we arranged to meet for dinner. I was taken at once with both herself and her fiancé. She shared my passion for the arts, but we had other things in common. Her father, my contemporary, had died eight years earlier. So we talked around these deaths, her father's and Patty's, in a thoughtful, wry way.

I studied my head in the mirror. As I visualised Patty's features superimposed upon my own, my mind filled with questions. What kind of blow – just what kind of blow – could have left her with part of a head? Lots of people had handled Patty's remains. They had recovered, assessed, catalogued and photographed them. They had applied themselves to identifying them. It bothered me that I had had no role in this. I had not been able to hold Patty's hand while she died, nor did I have the opportunity to pray by her remains. These were rituals: it did not seem appropriate that only strangers had been involved.

I went back to my desk. I wrote to Dr. Butt, the Chief Medical of Nova Scotia, and requested photographs.

I wandered around the house looking at our books. There were books in every room, thousands of them, spilling out from the packed shelves. How many had we read? How many would we read? Re-read? Life was shorter than the time it would take to read so many books. I thought of the house groaning, then dropping several feet into the ground, sunk by the tons of books.

I rifled the vast collection, culled a few hundred volumes. Nobody, not even Frank, would notice they were gone.

The telephone rang. I recognised Dr. Butt's sonorous voice at once. He was sending me the reports on Patty's remains. He hoped they would provide the information I needed. Any photographs would have to come through the Royal Canadian Mounted Police. This could be arranged, but I should read the reports first. 'Wait and see,' he advised.

A moving van drew up outside our house. Two men unloaded Patty's dressing tables, her *étagère*, then carton after carton, until my dining room was packed from floor to ceiling with cardboard boxes and furniture. As I opened each box, the smell of her apartment floated out: an acrid, New York, mothbally smell.

Unpacking Patty's possessions brought mixed emotions. I

felt like a thief who had stolen her belongings. But I was glad to be surrounded by things that reminded me of her. I made a conscious decision to use everything. We would drink out of the crystal and eat off the china. The idea of hoarding anything for a day that would never come haunted me.

I took out Patty's nightie. It still held her scent. Shaking the wrinkles out, I put it in a basket over my desk, where I could touch it.

I sat down. In front of me were a few rough pages of a story, 'Love', the story I had been writing the day Patty died. Painfully, the rest of the story bled out, word by word. At the end I wrote: *This story is dedicated to Patricia Wesley Eberhart, who perished on Flight 111, 2 September, 1998.* Then I cried and cried.

The reports arrived from Dr. Butt. The notes, accompanied by diagrams, were short and to the point. I was startled to discover that Patty's heart and lungs had been blown out of her torso. In a surprising way, this pleased me. I liked to think of her heart being billowed away by the sea. This was better than imagining it in a plastic bag in a refrigeration unit.

Patty's arms had been recovered and identified. Her fingers were intact, that is, until the pathology team had removed them. I read that panties (polyester floral) and the remains of tights had been found on her torso. I read also that there was a 'tampon in situ'. Then I stopped to consider briefly why it had been necessary to remove the shreds of clothing from a poor, flayed torso to determine whether or not it had been menstruating. Patty's ribs, pelvis and spine had multiple fractures. The description of her head, which included reference to an intact ear, was cursory, and left me with no clear image. Her legs and feet were not mentioned, so I concluded they had not been identified, and possibly not recovered.

I spent the morning with those brief notes, familiarising myself with every word, and every squiggle on the body diagrams.

❀

Late November. A friend's wife died from cancer. Ten days earlier, the doctor had drained a pint of fluid from her lungs. 'Good for the roses,' he had said to our friend. 'It's full of nutrients.'

❀

Christmas was coming. At the shopping centre, dozens of Christmas trees were propped against the wall. Andrew's ten-year-old eyes gleamed. The trees looked immense. Then I spotted a single, small tree in a tub. A living tree, plump, bushy and very much growing. 'I want to buy that little one in the pot,' I said.

Andrew looked horrified. 'I want a big one.'

'Then we can't buy it now, because a big one has to be tied to the car, but we can get the little one.' Reluctantly, Andrew agreed.

The tree looked bigger as soon as it was on top of Patty's antique bureau. By the time it was covered in white lights, it looked wonderful. Then Andrew remembered the angel ornaments that Patty had given us the year before. Together with scraps of red curly ribbon, the angels made the little tree glow.

Mother telephoned. Her voice was strong and happy. She had just arrived in Connecticut. After walking the last two miles, she had dumped her bag in front of the still-locked door and strode to the front of the deck. There, in the black cold of the night, she had lit a cigarette, gazing up through the smoke at the sky. Then her eye locked in on Venus, gleaming Venus, the bright star blazing in the firmament. Before her eyes, the star began to dip and spin, spin and dip, in a swirling, crystalline dance. 'It was Patricia,' my mother sang across the Atlantic, 'Patricia!'

⚜

Night-time in my car. I drove down Fosters Avenue, bearing left towards the turn that would take me home. Suddenly, I remembered Patty beside me in the passenger seat, at that exact place, a year earlier. Unaware of the feeder lane, she had thought I was going to slam into the car waiting at the lights. In terror she had screamed, 'Aren't you going to stop?'

The memory of her terror made my skin crawl.

⚜

A photograph fell out of a folder. It was Patty in her Pan Am uniform, sitting on the rim of the engine of a Boeing 707. She looked young and radiantly happy. On the back of the photograph she had written: Kennedy Airport, 1972. Not quite halfway through her life; a few short years into the career that would kill her.

⚜

23 December. I decided to go with Frank to his singing group. Before we left, I telephoned Mother. It was 7 pm in Dublin. There was no answer.

All that night, I was up and down to Andrew, who was feverish.

24 December. Andrew was sick. He was hopping mad too. Christmas Eve, for him, meant carol-singing at his much-loved violin teacher's home. Missing it was unbearable. We cuddled on the sofa and watched a video. Andrew's temperature was just under 103°.

Midnight, Christmas Eve. I was trying to get Mother on the telephone. I had been calling since 7 pm, Irish time. Now it was 7 pm in New York. As I listened to the ringing phone, I pictured my mother's bedroom: the old telephone, the chair heaped with papers, the jumble of boxes and folders on the floor. Where was she? Had she gone to a movie? Was she now eating her dinner somewhere downtown? The phone rang out, just as it had on the morning after Patty's death. I was filled with anxiety. Should I do something? What could I do?

I remembered the bitter tempest over the keys. And how, more recently, Mother had roared down the telephone, after I had complained about not getting through to her for a few days. If I called Adam, she would be livid. She was due, the next day, to have dinner at his house. Uneasily, I went to bed.

Hortense

Christmas Day. 7:45 am in New York, and Mother wasn't answering. This time, there could be no explanation, no Delta volunteers in the lobby. It was obvious to me that Mother was dead. I tried her number every 15 minutes. At 9:15 am, I phoned Adam.

He was sunny. 'She's probably gone for a walk,' he said.

At dawn on Christmas morning?

'I'll look into it,' he said.

Had it been her heart? Had she died in her recliner chair, surrounded by newspapers?

The day passed in a haze. We all began to cough and sneeze with a version of Andrew's bug. I put dinner on the table. As we prepared to eat, I asked that we remember the previous Christmas, when Patty and Mother had been with us. Then we pulled our Christmas crackers and put on the paper hats.

Another call revealed that Adam had gone to Mother's building with a locksmith. Soon I was speaking to the doormen, giving permission for the apartment to be broken into.

I waited.

When the phone rang, my husband passed the receiver to me. 'Your Mom's OK,' Adam said, 'but she needs some help.'

My mother's voice, loud and elated, burst through the receiver. 'No ambulance,' I heard her say. 'I won't go in any ambulance.'

Many hours later, Adam telephoned again. With the help of a locksmith, he said, he had broken into the apartment, where he found my mother on the bathroom floor. At the hospital she had been diagnosed with a broken hip, and a heart rate that was 'over the moon'. The plan was to stabilise her condition, then to operate. Adam had arranged both surgeon and cardiologist.

I didn't know how to thank him. 'How did you persuade my mother about the ambulance?'

'Simple,' he said. 'I just told her that the paramedics were my friends.' It was late, the witching hours in Dublin, but it was still Christmas night in New York. I thought about Adam's family and guests, and how disappointed they must have been without him.

The next day, I pieced together what had happened. Mother had fallen while trying to save a stack of papers from slithering off a table. She had tumbled over the junk she kept on the floor. 'I got up then,' she told me, 'but walking hurt. So I sat down, my behind facing the way I wanted to go. I put my hands behind me, and lifted. The phone began to ring. I knew it was the lawyer returning a call, but I couldn't get there in time.' Eventually, she succeeded in bottom-crawling to the phone, and calling the lawyer. 'I told him I'd had a bad fall,' she said, 'but that I'd be alright.' After hanging up, she dragged herself to the bathroom.

'Why the bathroom?'

'In case I wanted to throw up.' But once in the bathroom, she could move no farther. 'I got weaker and weaker,' she said. 'I was as miserable as a person could get.' She described the cold, and the indignity of peeing all over herself. 'But when the sun came out,' she added, 'it made the floor warm. Then I started enjoying it.'

Two-and-a-half days and nights, I thought. Or maybe even three. It was a wonder she was alive.

I phoned the lawyer at home. 'When did you last talk with my mother?' I asked.

'Monday,' he said, 'or maybe Tuesday. No,' he said, more certainly, 'it was Wednesday.' I told him what had happened. He was taken aback. Gradually he recalled their conversation. 'She had trouble getting to the phone, she said, because she had fallen. I said, "Get a cab. Go to the hospital." She kept saying, "Ouch! Ouch!" She was quite lucid.' The lawyer's voice reflected his distress. 'I should have done something.'

I reassured him. 'She thought she was in control. She made a conscious decision that she knew what she was doing.'

The lawyer had confirmed that Mother had spoken with him *after* her fall. My mother's pig-headedness appalled me. She had dragged herself to the phone and said she was hurting, then refused to do anything about it. And yet this was consistent with the woman I knew, determined not to relinquish her autonomy in any way. Mother believed she could deal with anything.

Various other narrative strands had been woven through Mother's account. One concerned her hospital room-mate. 'A spy,' my mother said. 'An eavesdropper! A lunatic!' The hospital wasn't much better. 'Something fierce is going on here,' she cried. 'I am a prisoner in this room. You don't suppose that they're aliens from outer space?' She pleaded with me to do something about the aliens, but I had to be careful: her phone was being tapped. Mine was too. So I should go to a pay phone and ring Adam. 'However,' she added, 'I'm telling you now that you won't be able to get him. But don't give up if you care about me.'

I did telephone Adam. 'My mother is saying some pretty strange things. Do you think she's all right?'

Adam was unfazed. 'She's wired on painkillers, and other drugs they're using to stabilise her condition.'

❧

I had to go back to New York, perhaps for another month. At once, I advertised for a minder for Andrew. I got no response. I phoned other mothers in his school. Only one said she would think about it.

Andrew became extravagantly unhappy when he heard. Hadn't I just been away? Didn't Mums belong with their kids? Beneath this truculence, Andrew's real terror seeped out. 'No way are you getting into any airplane,' he yelled, 'and that's final!'

I was gloomy about going myself. As always, my heart sank at the prospect of the cluttered, uncomfortable apartment. Mother, I knew, would be a taxing patient. And I didn't want to leave my family.

⁂

From her hospital bed, Mother was giving me a detailed account of how she managed drinking water in her apartment. She ran the tap in a particular way. She filled plastic bottles at the optimum moment. She stored them in the fridge, then decanted them in an orderly fashion.

'Really?' I said politely.

My mother took great umbrage at this innocent interjection. 'Shut up!' she fumed. 'Can't you let me finish just one fucking little thing?'

⁂

The lawyer phoned, after visiting Mother. She had been full of talk: about doctors who would operate only when they thought no one would sue them; and the punch of what she was getting through her drip. 'Magnesium,' she had said. 'I love magnesium.' She had also used pen-and-paper for part of their conversation to thwart the malignant eavesdropper, her room-mate.

The operation took place that evening. A few days later, Mother told me a story. 'I had just come out of surgery. The drugs made me feel that I was tied down. Next to me, there was a nice young woman having a baby. She was in difficulty. "Come on, dear," they said to her. "Please, dear." But she died. The baby lived, but she died. They wheeled her away.'

Adam told me not to come to New York until Mother was out of the hospital. In the meantime, he went to see her every day. His saintliness delighted my mother. 'He's so nice,' she said. 'So classy, so high-minded. He's made such a difference in our lives.' Adam brought her fruit, her reading glasses and a transistor radio with ear phones. He took away her washing. He brought a focus to her day, and he made her laugh.

The fact that Adam was tall, handsome and moustachioed was not wasted on her either. 'He makes me think that I'm the sweetest, most beloved person he has ever known. You don't know what it's like for an old bag like me to have such a charming young companion.'

Slowly, painfully, my mother began to take a few steps. 'If I were being paid for this,' she said, 'I'd be fired.'

The woman who had promised to look after Andrew changed her mind. I readvertised the position. I made an effort to take pleasure in the ordinary things of life: snowdrops blossoming; the shifting colours in the sky.

One day, as I was reading *The Irish Times*, a death notice caught my eye. Was it possible that the brother of a child in Andrew's class had died? Tragically, it proved true. For no apparent reason, a young boy had died in his sleep.

I went with Andrew to the funeral. The church was filled with children. At the beginning of the service, the mother addressed the congregation. At the end, the father did the same. Each word they said was filled with courage, dignity, serenity. Their response to this tragedy of tragedies was bracing. It put my own situation into a clearer perspective. What did I have to worry about, so long as my children woke up in the mornings?

I was on the telephone with Mother. 'Don't throw away my old newspapers!' she roared, apparently at a hospital cleaner.

'Oh, you're feeling better,' I said.

Four months had passed since Patty's death. Correspondence from Nova Scotia suggested that the recovery of human remains was complete. In a series of phone calls and faxes, I made arrangements for Patty's cremation.

Mother had been moved to the Rusk Institute, a centre for rehabilitation. In trying to find her, I caught up with Adam. 'Do you mind,' he asked, 'if I have her apartment painted?'

Did I mind? If the apartment was painted for the first time in twenty-five years?

Mother was telling me how she had had her operation on Christmas Day.

'No you didn't, Mom,' I said good-humouredly. 'On Christmas Day Adam broke into your apartment. They couldn't operate until your heart had calmed down.'

'I was there,' Mother growled. 'I know when my operation was. If Patricia were alive, she would tell you what was what.'

My plans fell into place. My mother-in-law's carer offered to mind Andrew with the help of her elderly parents.

But on the morning of my flight, Andrew woke up howling. 'I don't want you to go,' he bawled. Nothing would console him. He refused to say goodbye. Frank had to carry him out kicking to the car. Some thirteen hours later, I entered Mother's apartment in mid-town Manhattan with my key. Inside it was busy: three men were painting the living room; an electrician was tinkering; and the hall was packed with furniture. I went straight to the telephone. Andrew picked up on the first ring, and when he heard me, he whooped with joy.

I looked around my mother's apartment. Not only was the living room being painted. The kitchen was also different. The dense forest of accumulated detritus was gone. The surface of the counter gleamed, a sight I had not seen for a couple of decades. Cautiously I opened the refrigerator. Gone were all the mouldering take-aways, the shrivelled fruits and half-filled glasses of greenish liquid. And someone had shopped for staples: milk; bread; orange juice; pasta sauce; cheese; cold cuts; apples. For once, the food in my mother's fridge looked edible. The same hand had added a pleasing touch to the squalor of the second bedroom. There, wedged into a space, was a vase of fresh flowers.

Adam had been making magic.

I left the workmen and strode downtown. The irresistible vitality of First Avenue caught hold of me. Throughout the turmoil, I had forgotten what a very exciting place Manhattan is. With a thrill, I passed the towering sweep of the United Nations, remembering how my teenaged-self had sped this way to my best friend's apartment. It was all a long time ago, and yet, here I was, still alive and filled with joy. What had I known in my teens, when I took everything for granted, when I hadn't understood what a miracle the vast, turbulent city was?

At the Rusk Institute the lift whisked me to my mother's floor. I saw her at once, an aged figure in a wheelchair, framed theatrically by the verticals of a doorway. For a few moments I studied her ancient face, her fragility, her sadness. Beauty seemed to flow from her: a white narcissus, delicate with age. She glowed like a Rembrandt, like a Vermeer.

Then I took a deep breath and went in.

At first she didn't believe that I was there, but soon she began to talk. For two hours I listened to an unstoppable flow. She rode roughshod through the shortcomings of hospital, staff and her room-mates, then chirped sweetly about the wonders of Adam. Returning to the Furies, she recounted in venomous detail a visit from her least favourite niece. Eventually, she moved on to herself. 'The boys followed me like honey,' she recollected. 'I was a babe when I was young, but you weren't. And neither was your sister.' The memory of this youthful star quality made my mother happy indeed.

I walked uptown under the velvety sky. In the peaceful apartment, I ate at my little desk, feeling hopeful that everything would work out.

At dawn, I tackled mother's bathroom, the scene of her ordeal. The small room was jammed. Stuff was hanging from, and balanced on, the towel rails. Jars and bottles were heaped on the floor. Clutter covered the sink and the whatnot. When I opened the medicine chest, a bottle crashed into the sink. Back and forth I traipsed to the refuse room. Since Mother's hair care consisted of visits to a gentleman's barber, she would never miss the home perms and colouring kits, the hairnets, curlers and split-end repair sets, the hairsprays, some of which were so old that they didn't have bar codes.

After I had thrown out almost everything, a modest array of fresh, useful items remained, everything within reach. Then I turned to the hall. There, the mess was formidable, and, for the moment, all I did was clear a path, so that my mother might pass through with her walker.

In the Rusk Institute, I read through the leaflets about hip patients. There was a lot to learn fast. Dislocation of the new joint was a serious risk. To avoid it, various rules were to be followed religiously. My mother was never to raise her knee higher than her hip joint; she must sit in a chair with an elevated seat; she must never cross her legs or ankles; she was not to bend over; she must sleep on her back; and she wasn't to use the bath. Then, there was advice about installing grab rails in the bathrooms.

My mind was busy. Were there any suitable chairs in the apartment? And what about her bed? It was on castors, so what could I do to stop it from rolling away? And the toilets. There was one invalid toilet already in the hall of the apartment. Another needed to be ordered and delivered promptly, and a shower chair too. And just how was I to go about installing grab rails?

My mother interrupted my planning. 'You'd better take these home with you now,' she said. She started handing me things: plastic cutlery; sachets of salt, pepper and sugar; containers of juice and water; a wad of napkins – all retrieved from her meals. Exactly the kind of rubbish that Adam had cleared away from her kitchen.

My heart sank. 'What am I supposed to do with this stuff?'

'Take it home,' she said. 'Now! And all these newspapers. You won't be able to carry them tomorrow.'

'Mom!'

'Do what I say,' she hissed, 'and stop acting like a shithead. And this,' she said, pointing to a large, sturdy cushion, for raising the level of a chair. 'And these,' about a stack of empty bags.

'Isn't there anything you're going to throw out?'

'Of course.' Triumphantly, she held up a gleaming new paperback. Her least favourite niece had brought it as a present.

I staggered out of the Rusk Institute under a heavy load, all of it rubbish, except for the booster cushion. My mother had been so unpleasant that I didn't dare dump the lot into the nearest garbage can. What would I do if she asked for it when she got home? I boarded a crowded bus. Looking out the steamed-up window, I was envious of the pedestrians. I wanted to be walking myself, not weighed down with junk in a stuffy, smelly bus. My optimism was fading. Apparently, Mother was going to resist change every inch of the way.

I spent the evening cleaning up after the workmen; organising a chair; wedging my mother's bed; installing the invalid toilet; lifting carpets; further clearing a path. For a couple of hours, I slept like a dead person. I was back at the Rusk by 8:30 am. There, a young physiotherapist gave my mother and me a crash course in descending the stairs, six of which had to be negotiated if she was to get back into her building.

In the ward, a nurse handed me some papers. 'Sign,' she barked.

I began to read the papers.

'What's the problem?'

'I just want to read what I'm signing.'

The nurse was none too pleased by this formality. The papers detailed Mother's medical history and her medications. I asked a question about the drugs.

'I don't know,' the nurse shrugged. 'Her drugs have nothing to do with me.' Eventually I signed, after pointing out that my mother's operation had taken place on 28 December, not 25 December as the papers stated.

'No, no, no,' Mother protested loudly. 'My operation was on Christmas Day.'

We began to wait. Other patients were leaving. They were being taken away by helpers to ambulettes, small vehicles for transporting the infirm. 'By the way,' the disagreeable nurse said in passing, 'your mother refused to order an ambulette.'

I looked at my mother in her hospital wheelchair.

'I came here in one,' she said. 'They tied me down. It was so humiliating. And they charge you an arm and a leg for the privilege.'

'So how do you expect to get home?'

'On the First Avenue bus.'

She couldn't be serious. She could scarcely stand up and sit down.

'It had better be a taxi then.'

'Only if you pay,' Mother said.

We waited in the corridor for assistance, Mother in the wheelchair, and me with her walker, cane and the morning's supply of leftover breakfast, paper napkins and plastic cutlery. The young physiotherapist passed by. 'What are you still doing here?' he asked. When he heard, he wheeled my mother into the lift.

'Aren't you going to get a jacket?' I said.

He shrugged insouciantly. His tunic was short-sleeved. We made our way against the January blast up to First Avenue. All the cabs were full. The therapist began to shake with cold. Finally a taxi stopped. We helped Mother stand. We turned her backwards and lowered her into the seat. Then, patiently, the therapist coaxed and lifted her limbs inside.

'Please go very slowly,' I said to the cabbie. And slowly he went, weaving to avoid the potholes.

'Oh my God,' my mother said anxiously. 'I think I'm going to—'

'What?' I said.

A terrible smell filled the cab. I looked at Mother in alarm. She shook her head slightly, but I could not share her relief. I had other worries. How was I going to get her out of the taxi without raising her knee above her hip? When we stopped outside the building, the young cabbie unexpectedly jumped out to help. Together we eased the bad leg out, swung her halfway around. We lifted the other leg, rotating her the rest of the way. Then, with the walker in front, we hoisted her onto her feet. The taxi fare was just under five dollars. I handed the cabbie ten. If my mother had noticed, she would have had a fit.

To cross the street, I held her cane against the traffic. We got to the footpath, then inched our way to the stairs. Standing beneath her, holding her, I coached her as the physiotherapist had shown me. 'Bad foot. Good foot. Cane.' Three times for three stairs. Then my mother used the walker to negotiate the short distance before the next set of stairs. We repeated the cycle. 'Bad foot. Good foot. Cane.'

Miraculously, we were down safely. Mother was grey with fatigue. She shuffled across the foyer. She could scarcely lift the walker as she dragged herself onto the lift. Upstairs the length of the corridor stretched ahead, but the familiar ambience cheered her. For a while she just stood there, peering at the green-grey gloom. 'I thought I'd never see this again,' she said. And then I was inside my mother's head. I shared her astonishment, her wonder at the familiar. For four weeks, she had been living a nightmare of pain, handicap, strange faces and places. Before the hospital, she had endured two-and-a-half days on the floor in a puddle of her own urine. And three months before that, her daughter had been smashed into bits in a plane crash. Of late, her life had been full of surprises, all of them lousy. So why would the ordinary not look terrific?

It took another fifteen minutes for us to get down the corridor.

At last my mother was home. The workmen had gone. The drop cloths had vanished, the furniture had been put back in place. My mother was sitting in an appropriate chair, in a freshly painted living room. A foam wedge was lodged between her legs to keep her hip correctly aligned. A glass of juice was on the table beside her. I was grateful that we had made it without further disaster. However, I couldn't help but think how an ambulette and its crew would have achieved the same result with considerably less wear-and-tear on us both.

I went out again with Mother's prescriptions. Back upstairs, I drew up a chart for her medications. It was a fiddly business. One drug had to be taken twice a day, another three times. Certain drugs could not be taken with others. Some went with food; others with an empty stomach. The dosage of one varied on alternate days. There was even a drug that was meant to be taken in the middle of the night, as close to 2 am as possible. Allowing for all the requirements, two dozen tablets would have to be administered in the course of every 24 hours at eight different intervals, a daunting prospect.

The telephone rang. It was the nurse in charge of Mother's rehabilitation programme, who made an appointment for the morning. Immediately afterwards, the homecare agency telephoned with the name of a home help, who would work four hours, five days a week.

'No,' Mother said. 'I don't want these people. Tell them not to come.'

'Just the nurse, Mom. He has to see how you are.'

'I'm fine,' she said, 'and I don't want his dirty feet in my apartment.'

The prescriptions had to be collected. Mother looked sleepy. The thought of her dozing off and slipping onto the floor was alarming.

'Why don't you lie down while I'm out?'

'No.'

I sighed. 'Promise you won't go to sleep in that chair.'

'What a pessimist you are, Ivy. I didn't teach you to be a pessimist. I won't go to sleep in the chair. And get dinner while you're out. I want a pot pie from Kentucky Fried Chicken.'

The drugs came to more than a hundred dollars. The pharmacist scowled. 'Does your mother have insurance?'

'She's on Medicare. She's eighty years old.'

'I've allowed for that. I meant secondary insurance. Does your mother have secondary insurance?'

I shook my head. My mother believed neither in doctors nor in being ill.

With the drugs in my bag, I hiked up Second Avenue to Kentucky Fried Chicken. The fried food smell made me queasy. I had wanted to make a simple dinner of fresh foods, but Mother wouldn't hear of it. My preferences didn't interest her.

As I entered the apartment, I heard Mother snoring. I ran to the living room. She was slumped in the chair. The wedge had slipped from between her legs on to the floor, but at least she wasn't there with it.

It was time for Mother's bedtime medications. I explained what each pill was. This was a mistake, because Mother refused the iron pill. 'You may be anaemic,' I said. 'Why don't you just take it, and we'll ask the nurse about it tomorrow?'

'No.'

'Doctor's orders.'

'Don't be such a jerk. I brought you up to be a princess, so why are you such a jerk?'

The ancient bed sagged as my mother sat down, but the wedged wheels held. With a heave, she pulled up her left leg, then her right. I inserted the wedge between her legs, to hold the hip correctly. I covered her, leaving the walker aligned by the bed. In a minute, she was asleep.

In my sleep, I heard Mother stir. I flew in. I helped her pull herself up, get her legs down, lift her to her feet. With the walker, she started towards the bathroom. 'Oh my God,' she said. She was leaking a bit, but she made it. I was on the floor, mopping up, when she began to scream.

'What's the matter?'

'How dare you,' Mother raged. 'How dare you!'

'What have I done?'

'Your feet!' Mother bellowed.

I looked at my feet, then at Mother, enthroned on the toilet. She was livid because my feet were bare. We were not allowed to be barefoot in the apartment. So I retrieved my socks. Then I finished mopping the urine off the floor, returned her to bed, reorganised the wedge and blanket. The clock read a quarter past midnight.

I got up to her three more times that night. I even managed the 2 am meds. And I never again forgot my socks.

By 7 am I was at the kitchen table, which my mother used as a desk. It was a tangle of mail, financial papers, newspaper clippings and paper napkins. I sorted and tidied, with an eye out for anything connected with taxes. By 9 am, it was done, and I was serving breakfast in her bedroom.

The nurse was named Scipio. This historical association did something to allay my mother's hostility to his dirty feet. Besides he was the right sex. My mother prefers men to women, and Scipio provided a welcome, masculine audience. She told him what she knew about the Roman Empire, before moving on to the plane crash, and its role in her own accident. Then she suggested that he read *The Lives of the Cell*, a collection of scientific essays by Dr. Lewis Thomas.

Scipio made a note of the title. When he was able to get a word in edgeways, he outlined the nine-week programme for my mother's rehabilitation. In addition to himself, there would be two physiotherapists, a home help and someone to monitor her blood. He then examined her, inspecting the long, angry scar that stretched from her abdomen to the middle of her thigh. Her calves and ankles had begun to swell.

'Oedema,' he pronounced. 'You should be wearing support stockings.'

'No,' Mother said. 'They strangle your toes.'

'Get them properly fitted. They will keep the swelling down.' But she would not hear of them.

In the kitchen, Scipio was examining my mother's medications. 'A strong-minded lady,' he said to me.

⁕

We finished our take-away dinner. I tidied up, then scribbled an affectionate postcard. 'I'm just going to pop this into the mailbox,' I said. 'I'll be back in ten minutes.'

'What is it?'

'A postcard for Andrew.'

'Mail it tomorrow.'

'I promised him a postcard every day.'

'You're not going anywhere. You're here to help me.'

None too happily, I sat down.

⁕

I put Mother to bed. Then I tackled some of the piles in the hallway. For once, I used the refuse room nearest the apartment. Since Mother could scarcely walk, let alone manage the door, she couldn't check up on me.

After a while, I looked in on her. Her glasses were on, and her false teeth bared, but she was definitely asleep. So I nipped downstairs with Andrew's postcard.

Throughout the night, Mother roared and groaned, but she didn't wake up. All I could do was speak soothingly, stroke her head, and replace the dislodged wedge between her legs.

She was eating breakfast in her room when the doorbell rang. It was a sallow young woman, swathed messily in layers of clothing. In broken English, she explained that she was from the home help agency.

Why had she arrived an hour before the appointed time?

'What's your name?' I asked.

It took her a while to understand the question. When she finally answered, it was not with the name that the agency had given me. There was dirt under her fingernails, she had difficulties communicating, and her manner was abrupt. I didn't see how she could help my mother. When I told her so, she demanded to use the phone.

I shook my head. The only phone was in Mother's bedroom, and she would have been furious.

The doorbell rang again. Mr. Zahid. He had come to take Mother's blood.

'Dracula,' she said, offering him her arm.

At the door, a heavy-set man stood clutching an invalid toilet and a shower chair.

'Thank you,' I said.

'Wait,' he said. 'I need your credit card number.'

'No you don't. I supplied it when ordering.'

'Where's your telephone?' he demanded. Unwillingly, I let him call his company. Mother left him in no doubt as to what she thought of him. And of me.

'Adam is right,' I said to Mother. 'You need another phone extension.'

'Don't be stupid,' she replied.

I got the wrench and removed the seat from the toilet. Then I installed the invalid version. What wonderful skills I was developing.

Mother's energies were consumed by sitting, standing, taking a few steps, lying down and getting up. She couldn't move without the walker, which kept both hands occupied.

I remembered Patty's pinafore, part of her old Pan Am uniform, which was all enormous pockets. I slipped it over Mother's head. Now, she could keep what she needed handy.

❦

Scipio listened to my account of the home help. He shook his head. 'You're entitled to have a home help that you like,' he said. He got on the phone to the agency, and dressed them down. 'And send someone who speaks English,' he concluded. 'Mrs. Eberhart speaks English.'

Scipio measured my mother's still swelling calves and ankles. 'Support stockings,' he said again.

'Never.'

In the kitchen, he went through the medications, spotting a discrepancy with the warfarin: the tablets were of one strength, but the prescription indicated another. 'I'm cutting them,' I explained. 'Mother's instructions. She already had warfarin and refused to buy more.'

Scipio shook his head, pointing out that the dosage of warfarin was fine-tuned, and that a couple of grains either way could cause problems.

'There's no point in telling me,' I said. 'It's the patient you need to persuade.'

❦

Mother was searching the bookcase. 'What have you done with it?' she demanded. Her tone of voice was terrifying. 'You've thrown it out. Jesus Christ, you've thrown it out.'

'What are you talking about?'

'My notebook!'

'I haven't thrown out your notebook.'

'I know you have. You throw everything out.' Mother was out of control. Her eyes were flaming, her mouth twisted into a grimace. 'You're full of shit. You think you can take over. Well I've got different plans!'

'What does the notebook look like?'

'It's green. Any shithead would know it was green.'

'What was in it?'

'Stock records.'

'I wouldn't have thrown anything like that out.'

'Of course you would, you're such a jerk. Jesus.' She banged her walker against the floor and tore at the bookshelves. Anxiously, I scanned the dense, over-packed shelves, with no idea of what I was looking for.

Suddenly, she found it. Without apology, she stumped off, slowly and painfully, with the notebook in a pocket of the new pinafore.

⚜

Clarence, one of Adam's workmen, came to install grab rails. Mother was lying down.

'You don't mind if I let Clarence into your bathroom?'

'Whatever you say,' she mumbled.

Mother snored peacefully throughout the drilling. When Clarence was finished, I cleaned, wiping up the fine grit with moist rags. Mother did not believe in vacuuming the apartment. Instead, she used a dust mop, which she then picked clean over a plastic bag. It was neither hygienic nor efficient, but with patience it worked.

Clarence was gone when Mother awoke. At once, she was fractious because I hadn't gone to the bank.

'How could I go to the bank and leave Clarence in the apartment with you asleep?'

'Your sister,' Mother sneered, 'would have found a way.'

Mother wanted another pot pie. It was a joy to be out. I ducked into a pretty shop, which traded in ornamental objects. For five minutes, I dabbled in colour and texture. I touched things: the polished wood of a box; the rough silk of a sea-green throw. Then I detoured to a salad bar, before plodding onwards to Kentucky Fried Chicken.

Back in the apartment, I cut the lid off the carton of Mother's pot pie, arranging fruit in a colourful border around the crust. I knew better than to serve it on a plate. She hated dirtying dishes. 'Very artistic,' she said, as I put the carton in front of her. 'Didn't you get one for yourself?' Without comment, I crunched my way through green beans and broccoli. 'You just don't know what's good,' she said.

She ate the fruit and left half the pot pie. 'Put that in the refrigerator. I'll eat it for breakfast.'

⁕

Mother's feet were so swollen that they looked like pig's trotters. I filled a basin with water, eased her feet in, and let them soak. Then I dried them, cutting through the thick, overgrown toe-nails.

⁕

Pippa, the next home help, was fine. A couple of years my senior, she was neat, clean and smartly dressed. When Mother came in with her walker, Pippa stood up and said, 'Good morning, Mrs. Eberhart.'

Mother was pleased.

In half-an-hour, Pippa was helping her into the shower. I fled with a bag of laundry and some letters. My errands disposed of, I walked downtown quickly, almost running, twenty blocks down, then twenty blocks up. Back in the apartment, I took

Pippa aside, explaining about the mop and the plastic bag. I told her never to use Ajax on the ancient tub, and never to use Mother's bathroom. I showed her where things were, and suggested possibilities for lunch. Then cheerfully I got on with the Augean stables, delighted that there was now someone else to listen to Mother.

My plan was to clear everything off the floor, to make the apartment safer. Some things were easy to get rid of, like the basket filled with shower caps, soaps and disposable razors, or the dozens of too-small canvas shoes. Everything worth saving should be put away. However, since all wardrobes and drawers were packed to the point of explosion, I raided them, stripping away sacks of Mother's more outlandish garments. I knew that she wouldn't notice; she rarely went into the wardrobes or drawers anyway. The clothes that she actually wore had been kept in the shower, on the clothes rack that was now standing on the bedroom floor.

I had nearly finished the floor in the second bedroom. Near the wall where she had fallen, a couple of shopping bags were filled – inexplicably – with Russian newspapers. Thrown in among them was a tattered plastic bag. Its contents startled me: a wad of stock certificates. In their midst was a document that appeared to be the deeds to the apartment. When Mother was rested and cheerful, I showed her the bag and its contents.

'Where did you get that?'

'On the floor in my bedroom.'

'Put the bag on a hanger,' she said, 'and hang it in the wardrobe.'

'Wouldn't it be better to put the stock into your brokerage account?'

'Later. When I get better.'

'Why not do it now? We'll phone the broker, and I'll bring

him the certificates.' And that is what I did with $200,000 worth of stock. I also took the opportunity to give the deeds to the lawyer for safe-keeping. Then I hung the ratty bag, filled with copies of it all, in a bulging wardrobe.

⁂

The physical therapist had bleached blonde hair, a loud voice and 'very dirty shoes', according to my mother. She also had a regime of exercises that Mother was to do twice a day. The exercises were crucial, as a doctor had told me: 'How well your mother walks again depends on how she does those exercises.' Watching Mother struggle with them was agonising. She could barely lift her good leg into the air. To move the bad one required excruciating exertion. Each exercise had ten repetitions, and there were eight exercises. A Draconian total of 320 movements a day.

The walking itself was slow but comparatively easy. In the long corridor outside the apartment, the physical therapist encouraged my mother like a cheerleader. 'Atta girl, go for it!' she yelled. To the list of exercises, I added, 'Walk to lift, 3 times a day.'

The young occupational therapist was not so focused. She had a mane of curly hair and a mobile phone, both red rags to the bull in my mother. Most of the time, she succumbed to Mother's conversational onslaught, and spent her hour listening. When she did get on with the job, her exercises seemed to be about balance. She would have Mother lean against the wall for a count of ten, then straighten up. Then she would ask her to move items from shelf to shelf.

I added her exercises to the list, but they didn't pack the same punch as those of the physical therapist, and were often skipped.

The serious exercises took two hours and more out of every day. My mother did them neither willingly, nor independently. 'You are so cruel,' she would say. I tried to make it bearable by having her do what she could, while sitting. Then, when we came to the main sessions, there were fewer to get through.

My mother's hours were packed. The weekdays were punctuated with the therapists, the nurse, the blood man, the home help. The ordinary tasks of getting washed, fed and using the toilet, were time-consuming. The relentless exercises were exhausting. But dressing, at least, was easy. For Mother's underwear – that is, her lycra cycling shorts and tee-shirt – had always doubled as sleep and day wear. So, to dress in the overheated apartment, all she had to do was slip a loose skirt over her head.

Against her wishes, I ordered extensions for the telephone. The man from the phone company inspected the existing apparatus. It had been installed before he was born. He attempted to track the wiring. Frustrated, he left, coming back a couple of hours later with a buddy and a meter. They prowled around, eventually asking to open the linen closet.

They looked aghast at the overflowing shelves.

'All that stuff on the bottom has to come out.'

'OK.'

'And all that furniture, and all those bags and boxes, everything's got to come out from the wall. We don't move furniture.'

'OK.'

I began to shift the heavy furniture.

⚜

It was 9 am on a wet February morning. The telephone rang. It was my mother's least favourite niece. She was going to be in town for the day with her kids. They would be available at around 12:30 pm. Could we do lunch with them?

I demurred.

'Why not? I'd really like to see you.'

Outside, the rain was pouring down. In the living room, Adam's workmen were building a cabinet. The home help was due shortly. There were exercises to be done, and a new doctor was coming in the middle of the day, just around the time that this woman wanted 'to do lunch' with her children and my mother, an injured, elderly woman who, with assistance, could barely make it to the lift.

I demurred again, this time more firmly.

The new doctor was a geriatrician, tall, boyish and unassuming. My mother had taken a shine to him because he had cut her toenails on his rounds in the hospital. He parked himself on the edge of her sagging bed. He examined her swollen ankles and legs, which were peppered with red sores. 'You ought to wear support stockings, you know,' he said.

When she told him what she thought about support stockings, he didn't bat an eyelid. 'Alright, we'll try you on a diuretic.' Before leaving, he handed me a bundle of new prescriptions. Among them was a drug to counter the loss of calcium in my mother's bones. 'To be taken,' he said breezily, 'first thing in the morning.'

Throughout the day, the workmen had been busy with the cabinet, a desk with built-in storage. In the late afternoon, an altercation broke out between the men and a supervisor from the building. Workmen, it seemed, were not allowed in the building after five o'clock. In dudgeon, the men slunk out, promising to reappear in the morning.

There were layers of plaster grit and sawdust over everything. The mop was not up to the task. So, with moist rags, I took to my hands and knees.

<hr/>

Mother was eating pizza with anchovies, while winding herself up with old stories about people she didn't like. Eventually, in an effort to lighten the conversation, I said how I loved going out with my women friends.

'You're boring me,' she replied regally. 'Real women prefer the company of men. I'm afraid you're just a horse's ass.' Then she crossed her legs.

'Hey, Mom, you're not supposed to cross your legs! You'll dislocate your hip.'

'At least I'm not smoking.'

Apparently, the threat of dislocation and its impact on her recovery didn't worry her at all.

❧

I sat with the new drugs, slotting them into the time chart. There were detailed instructions attached to the drug for calcium absorption. Not only was it meant to be taken first thing in the morning: she was neither to eat nor lie down for an hour afterwards, to avoid nausea and dizziness.

Not to eat or lie down? How could that be right for Mother, who found it a Herculean effort just to get out of bed and remain upright? The drug could also cause liver damage, and it had cost $70. Foolishly, I told her this. Instantly she was on the phone, depositing a richly worded statement on to the doctor's answering machine. Then she telephoned the drugstore, demanding that they take the drug back, becoming further enraged when they told her that it was against the law to do so.

It was a long time before I could calm her down and get her to bed.

My head was filled with gloomy thoughts. My mother undoubtedly needed help, but after two weeks there were already problems in Dublin, especially with Andrew. I was needed by my own family too. How could I be in two places at once?

By 7 am the next morning, I was in tears. While working on the piles in the hallway, I had found a floral bag that Patty had given Mother. At once ten months had vanished and we were in Connecticut, Patty and I, singing Happy Birthday to Mother. It was a spring morning, and her bedroom was bright. She twirled around in her lycra shorts and tee-shirt. I had marvelled: to be eighty years old and so lithe. It had been a happy moment, there, in Connecticut on a glorious April morning. So recently, and yet, so long ago.

I went to Mother's bedroom to see if she was awake. Her eyes fluttered open. 'I deplore those jerks,' she said haughtily, 'especially....' The names of those she loathed, streamed out of her mouth.

'Good morning, Mom,' I said cheerfully. 'Would you like an egg with your oatmeal?'

'Don't you have anything else?'

She settled upon Wheatena, which I was cooking when I heard her scream. I ran for the bedroom, fearful that she had fallen. She hadn't. She was at the sink, hacking the back of her neck with a razor, trying to dry-shave it. 'You're no use to me at all,' she yelled. 'Didn't I tell you I wanted a hair cut?'

'You said you wanted Wheatena.'

'I need a hair cut! You can't keep me waiting forever.'

I raced back to the kitchen, turned off the gas stove, grabbed some scissors and attempted to cut Mother's hair as she stood at the sink. Afterwards, she refused the Wheatena. I wasn't surprised. It wasn't properly cooked.

⚜

I was showing Pippa, the home help, where to find a bucket.

'Your mother,' she said kindly, 'is an interesting lady. She reminds me of my father. My father liked my older sister better than me. It was very painful. Whenever he talked, it was hurtful.'

I squeezed her hand. She had been around for only a few days, but already she had made her own observations. But what could I say? I could hardly tell her that Mother had treated my sister exactly as she was treating me.

I moved a table in the living room, to enable my mother to pass freely with her walker. 'It's not safe,' I said.

'Don't be such a jackass,' she snapped.

'Fine, but if you hurt yourself some more, I'm the one who'll have to clean up the mess.'

'Don't be so fucking selfish.'

❧

Adam whooshed into the apartment, his arms full of files for the new cabinet. His presence worked upon Mother like magic. Before my eyes she became focused, contented, even girlish.

He asked her how she was.

'Oh well,' she laughed merrily. 'You know how it is. My daughter is not much of a nurse.'

❧

Pippa was giving Mother a shower, so I nipped out to buy more files for the cabinet. Mission accomplished, I couldn't resist the temptation of a bookstore. Inside, I noticed a familiar, gorgeous coat. The woman wearing it was one of the few people I knew in New York, and she was aware that I was caring for my mother. Instinctively I shrunk back so she wouldn't see me.

Since my reaction surprised me, I tried to work it out. I had seen the woman only a few days earlier. As we had chatted on the bus, I had read something in her manner: that she wanted to keep her distance. I didn't blame her. By avoiding her, I was saving her from the cursory involvement that results from the most casual 'How are you?' Of course, I might have appeased her by saying things were great, but that wouldn't have been right for me.

Racing back towards Mother's apartment, I passed one little restaurant after another. An idea took hold of me. Pippa would be with my mother for another hour. Why shouldn't I stop and have lunch? For forty minutes, I luxuriated in being treated like an adult. It was a welcome change.

⚜

With the cabinet completed, and plenty of files, it was time to organise my mother's papers. The idea was to remove them from the heaps on her bedroom floor, and to file them conveniently in the cabinet. I got stuck into the task, but Mother became agitated. 'Stop immediately!' she thundered. Towering over my chair, she beat the air with a broker's statement. 'You're no use to me whatever, unless you can write the old names of my stocks on every single statement!'

Frightened, I looked at her. Why was she so angry? What she was demanding was unreasonable. If, for some reason, she needed an old name, it could be traced, but to write the names on every statement would be pointless. Yet she was furiously insistent. What was going on? The intimations that had been niggling at me grew more fixed. However, Mother was still on the warpath. 'How dare you touch my papers anyway!' she roared.

'Don't you remember?' I said quickly. 'Adam had this cabinet built for them, because you can't pick things up off the floor, but you can sit here and use these drawers.' I showed her the clearly labelled files, and how the papers she needed were at the front.

'Hah!' my mother scowled. 'You're stupid and you're careless. You'll pay the price.'

With the cabinet up and running, I could begin working on her taxes. My object was to gather the documentation for her accountant. The accountant had supplied me with a copy of her list from the previous year, and I worked from that. Important things were everywhere, in opened and unopened mail, in the messy piles throughout the apartment, among the documents that were now neatly filed in the cabinet. Systematically, I began to track things down, using the quiet time in the early morning before my mother woke up.

꧁♣꧂

On a cold February morning, I went, in the company of the lawyer, to see the attorneys who were handling the compensation case for Patty's death. The issues appeared to be two-sided. Had Patty been employed by Swissair or by Delta? Was it prudent to opt for the short haul of chasing Swissair, or the long haul of going after McDonnell Douglas or Boeing? Was it better for a judge or a jury to hear the case? Other topics nosed their way in: the possible causes of the crash, and some pending legislation to amend the Death on the High Seas Act, which the lawyers kept referring to as Doe-sha (a verbalisation of DOHSA). However, as far as I could see, no amount of polite conversation could disguise the bottom line, that it was all simply about a figure. In short, how much was Patty's life worth? The real conflict was between the lawyers. The brief for my very nice attorneys was to extract as high a figure as possible, while the defendants' lawyers had the opposite objective.

I wasn't sorry when the meeting was over.

Over lunch, the family lawyer made it clear that he understood my uneasiness. He offered a more constructive perspective. When you pursue the defendants for the highest damages, he explained, the payout encourages the industry to make changes to prevent similar accidents. By way of illustration, he cited the controversial insulating material kapton. If kapton was shown to have contributed to the crash of Swissair 111, then an immense payout would bring about the removal of kapton from the commercial fleet.

I understood what the lawyer was saying, but the facts of Patty's death remained so raw, so consuming, that abstractions meant little. Her death had been devastating, and nothing could change that.

When I got home, Pippa had left. Mother was asleep on her bed, the wedge lodged correctly between her legs. For a while, I listened to the alarming pattern of her breathing. Noisy inhalation followed exhalation until, without warning, both stopped. The silence stretched, then stretched to a frightening distance, until, with a violent snort, she would inhale roughly, and the pattern would start again.

Mother woke in a temper about a statement that I had been meant to copy for the accountant. 'Why haven't you done it? I'll tell you why! Because you went out this morning to have a good time.'

In fact, I had made the copy on my way to the lawyers. 'Here it is,' I said, and I showed her the shelf on the cabinet, where I was accumulating the documentation for her taxes. 'And the original is here,' I said, showing her the file which held all the statements in their correct order.

My mother stared. Somehow, as far as she was concerned, I hadn't copied the right document; or if I had, I had made a botch of it. She abused my shortcomings roundly, finally sneering, 'Well, you're not Patricia anyway. Go back to Ireland. You're no use to me. You don't know what you're doing.'

'All right. I'll go tomorrow. Solve it all yourself.' With dignity, I walked off to my bedroom.

Twenty minutes later, she clumped with her walker to my door. 'I'm sorry,' she said. I was astonished. Apologies were not my mother's thing. Perhaps she really believed that I would pack my bags and return to Ireland.

'Just don't yell at me,' I said. 'It's very bad for me.' I meant it. My mother's shouting had always upset me. Things were hard enough without her eruptions of abuse. To be drained by unnecessary rages and explosions seemed unfair.

But the next morning things got worse. From the early hours, I had been working on the taxes. At 8:30 am, I took her breakfast, then settled down to another few minutes of paperwork. Unexpectedly I heard her walker, banging with force off the floor. Soon she was standing over me, bull angry, firing out questions about bank accounts. I answered as lucidly as I could, but she became more and more furious. 'You're so stupid! You just don't get it. Everything was fine until you got here.' Then she pushed her walker aside and took off towards the kitchen.

Frozen with horror, I watched her, a silent, wobbling wraith, hands extended ominously for balance. I retrieved the walker and went after her.

'Come on, Mom, use the walker.'

'Go to hell!'

Filled with terrified despair, I began to weep.

'Cry-baby!' she taunted. 'Cry-baby.'

I left the walker at the mouth of the kitchen, aligned so that she must use it, or move it. Then I retreated. It was clear that I had to do something. She was out of control, but whatever I did or said seemed to make her worse. Yet I was responsible for her. If she fell again, or dislocated her hip, it would be my fault. The nurse had made it clear that she was to use her walker at all times; her insurance would not cover her if she didn't.

I telephoned Adam. 'I need to talk to someone. I need some advice.'

I put the whole story to him: the temper, the abusive language, the daredevil recklessness. 'She gets confused about numbers and finances, then projects it on to me. According to her, I've lost things and screwed her up entirely; whatever's wrong is my fault. And then her fury makes her dangerous to herself. I can sort out the taxes, and the lawyer can manage her financial affairs, but how can her physical needs be met when she's so hostile to having anyone around?'

Adam listened. 'I'll tell her that she needs somebody,' he said. Then he laughed.

As I walked back to the apartment, I was a jumble of guilt and relief. Telling Adam about my mother, warts and all, had been indiscreet, but I felt the lifting of a weight, as if the responsibility were no longer entirely mine. And I knew that she would listen to him with a respect that did not extend to me.

That afternoon, when I let in the physical therapist, I mentioned that my mother had taken off without her walker. Her eyes clouded. 'That's very dangerous. Her bones wouldn't be up to it yet.'

'Would you please have a word with her? And you don't have to say that I told on her. You can approach it obliquely, as a safety issue.'

The therapist nodded. She was a smart, hard-working lady, and I liked her. Whatever she said to my mother, it worked. There were no more incidents without the walker, until her bones were sufficiently knitted.

***

Downtown. In the lawyer's lift, a handsome, middle-aged man smiled at me. 'Do you think that what we do today will make any difference?'

'I doubt it.'

'By the way,' he said, 'they're manufacturing a new Barbie doll. It's called Divorced Barbie and it comes with all Ken's things.'

I laughed politely. As I got off the lift, it struck me that he had been flirting with me. Somehow, this cheered me up.

***

The occupational therapist's mobile phone rang. 'I'm with a client now,' she said, and promptly disconnected. 'That was my husband,' she told us, tossing her mane of blue black curls.

Mother scowled.

When the therapist left, my mother was still cross. 'She has some nerve. Her telephone shouldn't ring when she's with me. I'm paying her to be with me.'

'She's young,' I said. 'Besides, isn't it sweet that her husband keeps in touch?' The incident seemed trivial, and I forgot all about it.

I plodded on with her taxes. It was uncongenial, but doing it helped me to understand how her affairs could be managed, when – and if – I went back to my family. The days were not easy. At 7 am, with 15 hours looming ahead, I felt grim. By 2:30 pm, I consoled myself with having reached a halfway point. At 10 pm, with Mother in bed, I congratulated myself for surviving. Weekends were the most demanding. Without home help, all I could do was dash out for an errand.

However bleak my life appeared, hers was worse. One night, as I put the wedge between her legs, her rheumy eyes latched on to mine. 'I will get better, won't I?' she asked.

'Of course you will, Mom.'

'It's a sure thing, isn't it? All I have to do is sit it out?'

'That's right. You'll be yourself in no time.'

You'll be yourself in no time.

Five years earlier Mother and I had been sitting on the sundeck of her house in Connecticut. It was a gorgeous day, sunny with a caressing breeze. My mother had a bottle of vinegar. 'The skin on my feet has grown hard,' she said, 'but the vinegar softens it.'

Companionably, I slipped off my socks. So we sat in the sunshine, rubbing vinegar into our soles. As we rubbed, she reminisced about a rented house in New Hampshire, where her own mother had died forty years earlier. 'A beautiful place in the middle of nowhere. The nearest neighbours were across the lake. One morning, a bat flew in. Your father chased it with a broom, while your grandmother giggled like a girl, dancing about the kitchen. She was only sixty-two, not old like I am now. When the bat was gone, she sent me back to bed and brought me breakfast. Up and down the stairs she went, telling me how wonderful I was. She even gave me her beach shoes. "You'll need them for swimming," she said; "that beach is full of rocks." Later, when your father had taken you girls off for ice cream, she sat me down for a cup of tea. She began to pour. She didn't stop. The tea splashed over the rim of the cup, flooded the saucer, drenched the table.

'"Mother, what are you doing?" I said.

'Then I saw the foam at the corner of her mouth, the strange expression on her face, the ice in her eyes. Your grandmother dropped like a stone.'

My mother stopped her story there. She was finished with the vinegar. Her bones creaked as she straightened her leg. Her breath rasped, seasoned by decades of Camel cigarettes. She brushed away a tear. Then I understood that it wasn't her mother's death she was thinking about. It was her own.

14 February, the second anniversary of my father's death, and not six months since Patty's. Mother was clutching the Valentine I had given her. 'Eight copies, that's what I want!'

'Why?'

'To give them to people. To show them how funny you are.'

'Four copies?' I said hopefully.

'Eight, eight!'

I had eight copies made.

Pippa told me that I had just missed the nurse. Mother had been bending his ear with complaints about the occupational therapist, her frizzy hair and her mobile phone.

When the occupational therapist arrived, Mother launched into a direct verbal assault. The therapist did not listen for long. After a few futile attempts to cut across my mother's colourful verbiage, she announced her immediate departure.

'You're not leaving,' Mother said majestically. 'I am throwing you out.'

The therapist pushed away my soothing words. 'I have never been spoken to so rudely in my life.'

Mother was exultant. 'I like it when people hate me.'

I was demoralised, but not surprised. My mother had always thrived on confrontation, but this one seemed ill-judged. Mother was meant to be working with her helpers towards her own rehabilitation. Instead, she had agitated herself over a mobile phone and a hair-do. At the same time, I didn't think that the therapist had behaved well either. After all, my mother was an elderly patient who had undergone severe physical and emotional trauma. A more experienced professional would have defused the situation by apologising

Gloating, Mother telephoned the hospital. In dulcet tones, she voiced her complaint, beaming across the room. I didn't miss the subliminal message in her smile, for, on one level, her dismissal of the therapist was directed at me. I had made the mistake of not sharing my mother's outrage with the therapist. So, by throwing her out, she was showing me who was boss.

I wandered into a church on 72nd Street, where I prayed for the situation to move forward. I wanted to go home and see my family. I wanted to live my own life. As I prayed, Patty's words came into my head: 'You're not going to give me any trouble, are you, Mother?'

There was black comedy in it somewhere.

We were sitting in the living room. I was brooding about the occupational therapist. The episode summarised the difficulty of getting help for Mother: she didn't want anyone around; and people found close contact with her challenging.

'You look so unhappy,' Mother said breezily.

'I *am* unhappy.'

'*Tant pis pour toi.*'

We listened to the radio. The presenter was introducing Mahler's song cycle, *Das Lied von der Erde*. I looked at my watch. It was nearly 10 pm. 'Do you want to go to bed?' I asked. Mother shook her head. Mahler's cadences began to groan and grate out of the radio, making my skin crawl. Grumpily, I went off and brushed my teeth. If Mother wouldn't go to bed, I couldn't either. And if I didn't sleep, how could I do her tax work?

She turned up the volume. The yowling music racketed throughout the apartment.

The following morning I made an unpleasant discovery. Her tax return for the previous year was missing. If Mother went looking for it, there would be hysteria, violence, and the insistence that I had thrown it out. And over such a document, the explosion would be worse than anything I had experienced yet.

I fled to the bathroom and threw up.

Then I telephoned the lawyer, who was coming uptown to see us. I left a message on his machine, asking him to bring a photocopy of the missing return, if he had one. If the lawyer had a copy, I could substitute it, and pray that she wouldn't notice.

I was shaking with anxiety.

We had just sat down with the lawyer. My mother studied me and sighed theatrically. 'Why don't you fix yourself up? Why don't you comb your hair?'

The lawyer opened his briefcase. 'By the way,' he said, 'I've brought the copy you asked for.' My blood froze. But Mother was still absorbed by the deficiencies of my appearance, and I was able to slip the papers into my notebook. The meeting proceeded. The lawyer was efficient. By the time he left, the seeds for the orderly management of Mother's affairs had been sown.

I was exhausted, drained by fear. 'I'm going out,' I announced, and within seconds, I was gone. I knew there must be pedestrian access to the river a few blocks east. On a hunch, I turned down 71st Street. From York Avenue, I could see a metal walkover rising above the street. My pace quickened. At the first sight of the glint and swell of the water, I felt better. The illimitable stretch of the East River was a joy after the tensions of the apartment. The surging tide excited me. I turned south and walked, oblivious to the streaming traffic. I looked only left, at the bright, swirling river.

My mother's fixation on my hair was nothing new. From the time I was approaching puberty, she had insisted that my hair be cut short. I was to be a Danish princess, Mother proclaimed, and this was to be achieved by having a barber crop my hair. The idea of being a Danish princess was demoralising. It was the 1960s and I wanted to be Marianne Faithfull or Joan Baez. I fancied Carnaby Street, not Copenhagen, and short hair made me miserable. Matters came to a head on the afternoon before a school dance. My dream was to attract the attentions of one Jason Robards Jr., the handsome son of the movie star, who attended a nearby boys' school. But when I put on my glasses to assess my personal attractions, I was devastated. No way, I decided, would Jason Robards Jr. look twice at a girl who resembled a boy.

Contemplating the dance with growing desperation, I turned to a copy of *Cosmopolitan*, a periodical that Danish princesses didn't read. There, in a column of Beauty Tips, I read that cod liver oil applied to the eyelashes made them longer and thicker. Hope gleamed. Surely if thickening and lengthening could be forced upon the hairs of the eyelid, then what miraculous transformation might be attained by dousing the entire head? So I shelled out my nickels and dimes for the finest cod liver oil. Never before was cod liver oil applied to the scalp with such enthusiasm. I massaged and kneaded until every short hair on my head was saturated, until I resembled the slipperiest of seals. Then I sat back and waited for the miracle.

It occurred to me that I didn't know how long it would take. I checked the magazine again. It didn't say, nor was there anything about removing the oil. So I waited, peering into the mirror, but the length of my hair remained unchanged. Time was running out. No way could I go to the dance looking and smelling like a seal, so reluctantly I got into the shower and shampooed energetically. The rinse water glowed about my ankles like a rainbow.

In the darkened school auditorium, a group called The King James Version belted out *Wild Thing*. With the other wallflowers, I waited to be asked to dance, each terrible minute seeming an hour. Finally a shadowy figure loomed before me. Could it be? Was this shadow the longed-for Jason Robards Jr.? I couldn't tell because I wasn't wearing my glasses. At any rate, we bounced about on the dance floor, working up a sweat. Then, when the music slowed down, my partner latched, limpet-like, on to me. Please, I prayed, oh please let it be Jason Robards Jr. Suddenly I heard a mumbling in my ear. Oh, miraculous, I thought, was the young man asking for my phone number?

The mumbling repeated itself. This time, I picked out the words. 'Do you smell something?'

I inhaled. There was definitely a smell. It was something…

fishy. With growing horror, I watched the young man sniff again, then his eyes focused accusingly on my still oily hair.

I fled, smashing out through red doors into the grey streets of Manhattan, blinded with tears, determined to rebel against Danish princessdom forever. From that night, I refused the barber point-blankly. Just so, Mother dug in, relentlessly voicing her disapproval of my hair, often several times a day, in tones that ranged from anger to derision. And so it went on, a rooted disapproval, unmitigated by the passing of decades, a disapproval that survived a host of hairstyles and tragedies.

Mother's interest in managing the piles of mail that arrived daily did not increase. She neither winnowed out the junk mail, nor prioritised what required action. When she did decide that something was important, she would hide it, then forget what she had done with it. I tried to get her to put aside important papers for the lawyer. I got a box and a marker. 'Will I use his first or his last name?' I asked.

She looked at me unhappily.

'How about his initials then?'

'That's better,' she said. 'Never give anyone any information that they don't need.' Then I encouraged her to put statements and bills into the initialled box. But after a couple of weeks, the only correspondence that Mother put in independently was about renewing her opera subscription.

Walking regularly along the river helped me organise my thoughts. Mother's rehabilitation programme had another five weeks to run. After that, it seemed, she would still need assistance. Where was that help to come from? How could she be persuaded to accept it? As it was, I was the conduit for her helpers. How would Mother manage without me? She could neither go out nor shower independently. She could not bend over or tie her shoelaces. If she broke a glass, she wouldn't be able to clear it up. Even she must see that she needed help in the short term.

If Mother could be persuaded to accept the minimum of four hours daily, then what would she do for the rest of the day? As it was, I was on call most of the time. Yet she claimed that she could manage on her own.

It was time to put her theory to the test.

'You wouldn't mind if I spent more time out,' I said casually. But Mother did mind.

'You're punishing me.'

'You say that you don't need help. But you're not managing on your own if I'm always here.'

'You're oozing resentment,' she growled, 'and I just don't like it.'

Nonetheless I began to go out more, especially on weekends.

Whimsically, I went to the zoo in Central Park. The weather was icy but beautiful, and the zoo was enchanting. I lost myself in the Rain Forest habitat, wandering through the steaming exotica, lianas twisting over my head. The contrast between the wintry city and the sultry jungle could not have been greater. The morning disappeared in the luxury of new experience, none of it connected with my mother.

Back in the apartment, I broached the question of carers, by asking Mother what qualities she would value in an ideal helper. At once her face flooded with temper. Nothing could persuade her to apply herself to constructive thinking instead of negative emotions. She had not calmed down when the telephone rang an hour later. It was Frank, with Andrew, who was pining to talk with me.

'Hang up,' Mother demanded. 'I want to read to you. Now, while there's enough light to see.'

'I'm sorry, but I'm going to talk with my son. He's ten years old and he wants a chat before he goes to bed.' At that, Mother became very angry indeed, so I took the call in the bedroom.

The sound of crying woke me up. I followed it through the dark apartment to discover Mother in the bathroom with the door open. Tears were pouring down her wrinkled face. Her bony fingers clawed through the bum bag clipped to her waist, loosing a muddle of worn newspaper clippings, envelopes, money.

'What's the matter?'

'I can't find it.'

'What?'

'My Medicare card. I might as well be dead without my Medicare card.'

'We'll find it later. It's not worth crying about.'

A fresh shuddering of terrible tears. 'I killed her,' Mother wailed. 'I killed her.'

'Don't be silly. The cockpit was on fire. You didn't kill her.' Then I coaxed my mother, pleaded with her, attempted distraction. Nothing worked. Nothing would persuade her to do anything except weep for her dead daughter. She would not be consoled.

From some other world I looked down at this elderly woman weeping on a toilet, her surviving daughter helpless beside her. I saw the terrible failure of their relationship, the fact that one could offer no solace that the other would accept.

8:30 am. I was at the cabinet labouring over taxes. The thumping of the walker signalled Mother's approach. She stopped and bared her teeth. 'You're here to be helping me,' she complained. 'You shouldn't be doing your own work. You are the most selfish person in the world.'

'Since seven o'clock, I have been working on your taxes.'

'Horseshit!' she said, and clumped off.

'I don't want any funeral,' my mother said for the hundredth time. 'When I die, you're to telephone Cremations Direct.'

Cremations Direct? Could there possibly be such a place?

On Second Avenue I ran into Mary Miley, a woman who had taught me piano as a child. Recently, Mary had fallen and broken both ankles. Yet here she was, marching spryly along the footpath, and she was a couple of years older than Mother. Delighted to see me, Mary tugged me at once into a supermarket, because she couldn't hear over the roaring traffic. There, we chatted and joked. She was hurt that Mother wouldn't let her visit. 'A demon,' Mary commented, in her dry, Oklahoma accent. 'Hell on wheels, just like my own mother. And *she* didn't settle down until she had a stroke!'

When we said goodbye, my eyes followed her admiringly, a lively old lady, whose resilience was uplifting. I hit upon a plan. Why not invite Mary over to give my mother an image of recovery? After all, if Mary could walk well after two broken ankles, why couldn't she? Mother predictably refused. 'Don't we have enough dirty feet in the house, already?'

6 am, Sunday morning. I was dreaming in bed. My plane was touching down in Dublin. In moments, I would be flinging my arms around Frank and the boys. Mother came into my

bedroom. 'Wake up at once. I need the tablets that clean my dentures.'

'In the kitchen drawer,' I grunted sleepily.

'They're gone. You've got to go to the drugstore.'

I got up, grabbed a handful of fizzy tablets and gave them to her. Then I went back to bed.

By the time I got up again, she was asleep, presumably with sparkling dentures. I looked up Cremation in the Yellow Pages. There actually was a listing for Cremations Direct. How very New York, I thought. Taking out the dead like trash.

Having gotten rid of the occupational therapist, my mother was winding herself up about her other helpers. As we plodded up and down the corridor outside her apartment, Mother skewered their characters with words. One was a 'monster'. Another had 'plotted' with the occupational therapist against her. The lot of them were on a gravy train, going through the motions for easy benefits. Above all, they were listening, recording the minutiae of our lives, 'so as to use everything against us'. In a phrase, it was us against them, only Mother did not confine herself to a phrase. On and on she rambled in a vitriolic loop, her slaying of character undiminished by the immense demands that walking made upon her. I concentrated on her walking, directing her as the physical therapist had shown me. 'You don't listen enough,' my mother complained bitterly. 'You don't hear what I say.'

'I do listen.'

However, defending myself only made her angrier. What she really wanted was for me to endorse her vision of the healthcare workers as scheming opportunists, and this I was not able to do.

A raw February morning with a scouring cold that burnt my face.

Downtown, in the warmth of Barnes and Noble, I discovered a sale table of Books on Tape. Stretching out the luxury of choosing, I settled upon four. Then I headed back into the cold with a view to walking up Fifth Avenue. However the blast of arctic air blowing off Central Park proved to be too much. I ducked into the shelter of a side street, where everything I saw reminded me of Patty. Every shop, every restaurant brought back the past. My feet led me to the Alliance Française. I stood across the street, my eyes tensed upon the door. Could I will her to come out? My eyes burned fiercely. I stared and stared, but Patty did not, would not, could not come out. The tears on my cheeks were ice.

❧

Adam was on the telephone, speaking with Mother. Her eyes shone. For him, she was a debutante at a party. 'I wish I could get my daughter to go back to Ireland,' she said, with a dainty laugh.

I could hear Adam's voice through the receiver. He wanted to take me to lunch. 'Well I suppose I can let her go,' Mother said, 'just this once.'

❧

'Ivy?' Mother said, 'why don't you send Richard to stay with me?'

'Mmm.' I thought of my strapping nineteen-year-old son, umbilically connected to his computer. The idea of him hulking around the uncomfortable apartment – with neither computer nor watchable television – was comical. He would have no patience with Mother's moods. When she shouted at him, he would shout back, only louder.

'It would be good for him,' she said.

'Great idea. Unfortunately, he can't take the time off from college.'

That she had asked was a good sign. Subconsciously, anyway, she knew that she needed help, and that I had to go home.

It was dawn. For the last time, I checked through the documentation for my mother's taxes.

When Pippa arrived, I headed for the river. Briskly I strode beside the bright, heaving waters to 91st Street and back. I was ridiculously, ecstatically happy. The taxes were done, and for a few brief hours, I could do what I liked.

I sat across from Adam in the Café Heidelberg. Its ambience stirred the past, bringing images of my father. *Mein Gott! Schön sechs Jahr alt!* he had written on a German-made card for my sixth birthday. He had brought us to German restaurants, read us Grimms' fairytales, introduced us to the happy delights of German confectionary.

My immense, sunny cousin pulled me back into the present. He was saying that I should set a date for going home to my family. In theory, he was right. But physically, mentally and emotionally my mother was in an uproar. Bodily injured, she wasn't able to look after herself, but intellectually, she was warring against dependence. How could I leave before a framework for her care had been set in place? Yet since she was hostile to the very idea, we were at an impasse.

Adam proposed a meeting: Mother, me, himself and the lawyer. In such a circle, he thought, she was bound to see reason. He would also act, he said, as an immediate point of contact for her care workers. I was astonished. Adam ran a demanding business, and he had a family of his own. For some years he had been involved in the care of two other elderly relatives. Surely collecting old women was an eccentric hobby?

He grinned. 'Only relatives. I don't pick them up off the street.' Adam ordered another sauerbraten – to take away – for Mother, a magic act that delighted her.

Mother was being difficult with Nadia, a nurse practitioner, treating Nadia's every question with suspicion.

'Subtract seven from a hundred,' Nadia said.

'Ninety-three.'

'Good. Now please take seven from that again.'

'Nine,' my mother said, and knew at once that she was wrong. She talked her way through a second attempt. 'Eight,' she concluded. Nadia nodded, and moved on.

After Nadia left, Mother asked for her calculator. 'I'm going to work it out,' she said. 'If I memorise the answer, I'll know the next time.'

At night I was restless. Patty drifted in and out of my dreams, a vague, shadowy figure living somewhere in France. Then my father appeared more vividly, dwarf-sized, no larger than a six-year-old. Raving with Alzheimer's disease, he seemed unable to keep his trousers aloft. I woke, my mind fixed on Father. A particular episode came into my head. It was the summer of 1970. We were travelling together on the New Haven Railroad, headed north out of New York. As always, he was wearing a conservative suit, a white shirt, a mildly eccentric tie. Suddenly he put down his *Wall Street Journal* and asked, 'Have you ever sampled any of this marijuana stuff? I'd like to try some myself.'

As it happened, I was nineteen years old, and had tried pot once. I had been keen to sample the forbidden fruit, because my contemporaries all claimed to have done so, and I wished to appear up-to-date. Finding myself at a party where a water pipe was being circulated, I watched hawk-eyed, studying technique so that when my turn came, I would know what to do. The idea, I saw, was to suck in as much smoke as possible, inhaled in a staccato of short, deep puffs, letting it percolate your insides. Then you let it go, together with a mellifluous sigh that proclaimed benignity and ecstasy.

It looked simple enough, so I took the pipe, inhaled like a vacuum and severely scorched my lip. Repressing a scream, I managed a grimace of bliss. And when the pipe next came around, I waved it on, hoping that the rest of the company was too far gone to notice the blister rising on my sizzling lip.

I had been too humiliated to share this experience with anyone. But now, my own father wanted to talk about it, and on the New Haven Railroad. What's more, he even wanted to try some marijuana for himself. I was mortified that my father, who (I had assumed) believed grass to be the green stuff that grew on lawns, should want to discuss such a thing. It wasn't decent. After all, it was I who was the flower child in my granny dress and silly specs, and he was … well, he was my father. And there were things that fathers and daughters didn't talk about. So I steered the subject towards less intimate territory, and the opportunity passed.

Now, in the sleepless hours, I understood how I had blown it. Thirty years too late, I realised that there, on the New Haven Railroad, my father had reached out to me as a fellow adult – something that Mother still could not do – and that, irony of ironies, I had brushed him aside.

⁂

The day was good for late February. The sun was shining, and there wasn't any wind. 'Let's go up to the roof garden,' I said. I pulled a warm pair of leggings on to Mother. I helped her into my down coat that swathed her to the ankles. Then, she hobbled into the lift.

Soon, we were sitting in the sunshine, Manhattan's skyline around us, a stretch of blue over our heads. My mother looked about in wonder. More than two months had passed since her accident. For five weeks she had been struggling for mobility inside her apartment. Now, at last, she was out in the air, the bright light dancing on her cheeks.

In the evening, Mother talked about Patty and a cousin. As teenagers, the two girls had exchanged uncomplimentary notes about their parents. 'Your sister said, "My Mom always yells at my Dad." Your cousin replied, "My Dad always makes my Mom cry."'

My Mom always yells at my Dad. 'Fishing tomorrow!' I heard my father say, as he tapped his cap and grinned. We were in Connecticut; Patty was ten and I was six. My father woke us before dawn. We trotted beside him past sleeping cottages towards the marina. Engine purring, our small outboard glided past the deserted boardwalk, emerging in ghostly splendour with the sun. Once anchored on the blue-green waters, Father took out some slimy sandworms, dismembered them and impaled them wriggling on our hooks. We sat back and waited. Suddenly, his reel began to spin, as yards below us some unsuspecting creature seized the bait and began its run. After half-an-hour's struggle, my father's prize thrashed fighting through the surface of the water. It was a fine blackfish. 'Twelve pounds if an ounce!' he cried.

Like a hunter, he carried that blackfish home, photographed it, then gutted it for dinner. My mother, however, could not share his pleasure. For some reason we didn't understand, she picked up the fish and flung it at him. My father ducked. With horror, I watched that blackfish swim again, headless and tail-less, across the black and white tiles of the kitchen floor, leaving a slippery trail before skidding to a stop beneath the stove.

In the morning, I again mentioned homecare. Mother exploded. 'I don't need anyone! Look how well I managed when I had my skin cancer removed.'

'Look how well you managed when you broke your hip.'

This home truth enflamed her further.

'You let our differences show in front of other people,' she spat. 'It's low class. Don't do it, or I'll crack your skull in front of everybody.'

'Look, Mom,' I said, trying to sound reasonable. 'I can't stay here for ever. I have to go back to my family.'

'Hah!' she snorted. 'You can take your family and you can shove them up your behind.'

Mother went back to bed. When she woke, we did some exercises, after which she slept again. Lunch was followed by more exercises and another nap. In the late afternoon, we went for our walk in the corridor.

'Come on, let's do it again,' I said, after we had completed a length.

'Only as far as the lift.'

On the way back, I stopped at the fire door. 'Come on,' I said. 'We have to do stairs if you want to get out of this building.'

By the time we got back, she was exhausted. 'You made me walk the distance to 57th Street,' she complained.

I could appreciate how she found it interminable.

On Monday morning, I was astonished to hear Mother ask Pippa if she would stay on after the rehab period was over. Pippa seemed interested. However, her agency was shoddy, and although I liked her, she wasn't reliable: during her time with my mother, she invariably missed one day out of five, and the agency's substitutes had all been lousy. My mother needed a better service. Besides, she was soon back-pedalling furiously. 'I don't want to live unless I can be independent,' she wailed. 'What will I do with these people?'

'They will clean, shop and cook for you,' I said. But her face remained stony. Like Greta Garbo, she wanted to be alone.

Since Mother was due to see her cardiologist, we needed to practise getting out on the turbulent city streets. Once again, I bundled her up in my leggings and the light, warm coat. This time, we took the lift down to the lobby. The doorman's delight at seeing her was a spur. We sailed out to the stairs, which she conquered. On the street, she ploughed on, keen to prove herself. My grip on her arm was iron. By the corner, she was breathless, so we rested.

In his poem 'Canal Bank Walk', the Irish poet Patrick Kavanagh writes about how – after a long illness – the ordinary things in life appear with splendour. And so it was for my mother, standing at the corner of Second Avenue, marvelling at everything. After a while we turned to a nearby shop window, a dazzle of decorative objects. She feasted her eyes on each item, luxuriating in the words that she found to describe them.

After a while I asked, 'Shall we go back?'

She shook her head. So we moved on, slowly now, past the delicatessen and the locksmith, past the new sushi restaurant, past the mirror and frame shop, coming finally to the voluptuous display of flowers at the next corner. There, Mother searched her head for the name of every blossom, drinking in their colours and shapes. After a long time, we picked our way back.

'Tomorrow it's around the block,' I said. Glowing, Mother fell into bed and slept for two hours.

In the evening, she remembered how she had encouraged Patty to work for Pan Am. 'If I had known then what I know now about medicine,' Mother said, 'I would have put her on to a different track. Nurse Eberhart. She would be running New York Hospital by now. She would be president. President of New York Hospital. There would be no anti-white stuff going on. Patricia wouldn't permit it.'

'Dreamer,' I heard Patty saying.

The following morning, without comment, my mother took

an empty milk carton and opened the door, headed for the refuse room. Another first. For a moment, my heart fluttered, and not because I thought she wouldn't make it. What worried me was whether or not I had left anything in the refuse room that she would recognise. But when she returned, all she had was newspapers.

❧

It was nearly noon. With a worried look, Nadia was taking Mother's blood pressure for the third time. The blood pressure was way down, and her heart beat was up. What's more, there was a problem with her drugs. One of them, apparently, was not recommended in conjunction with another.

❧

James, the new occupational therapist, came for a final visit, bringing several strips of brightly coloured rubber. My mother was to stretch these, providing light resistance training for her arms and chest. She took to these bands with enthusiasm. They became the only exercise that she ever did independently.

As James rose to go, he smiled sweetly. 'I wish I could come back,' he said. 'You're so interesting.'

Mother was ecstatic.

If only I could find someone exactly like James for her homecare, I thought.

Pleased by James, Mother was glowing from ear to ear. She did not look like someone whose heartbeat was too fast, and whose blood pressure was too low. Bloody hell, I thought. 'Let's stick with our plan,' I said, 'and go around the block.'

'Let's.'

What if she keels over in the street? I thought. But it was a gorgeous day, bright and not too cold. And she needed to feel confident outdoors, if we were to keep her appointment with the cardiologist. 'You have to promise not to drop dead on this walk,' I said dryly. 'They'd probably take me to court.'

'Say that I made you do it. Say that I beat you with my cane.'

This time, aware of the challenge, she left the building at a snail's pace. At the florist's corner, we headed east. The street extended ahead, looking interminable. Never before had I realised how far it is between Manhattan's avenues. Every couple of paces Mother stopped, anatomising the frontage of each brownstone, calling my attention to cornices, stairs, foot-scrapers, a heap of logs. There were plenty of hazards: jostling teenagers and lively dogs. I held on to Mother for all I was worth. The only thing that mattered was that she was happy, revelling in being alive.

After a very long time we turned the corner. First Avenue was a bustle of shops and restaurants, and she inspected every one. Another corner, and we were on the final lap. The street was suffused with light. On our left, children shouted in the playground. On our right, a vine sprawled up some old brickwork. Incredibly, I heard birdsong.

'I hear it too,' Mother said. We peered upwards through the leaves. The vine was alive with sparrows, dozens of them, chirping in the sooty greenery, as the cross-town bus rumbled a few feet away.

We were home. Our odyssey had taken two hours, but nothing had gone wrong.

My mother was standing at the open cupboard in the kitchen. 'You really ought to clean this,' she complained. 'Just look at this schmutz.'

'Why don't you clean it yourself?' I said nicely.

Propelled by the confusion over Mother's medications, I emptied the public library of books about prescription drugs. While Mother slept, I read about drugs. It was sobering. Eight times a day, my mother was downing a cocktail of formidable substances. Potential side-effects included acute nausea, dizziness, liver damage, internal bleeding and confusion. My mind focused on the last. Was it possible that Mother's difficulty with numbers was being caused – or exacerbated – by prescription drugs?

⁂

The weather deteriorated. The falling sleet was whipped by the wind into ridges of ice on the footpath.

Mother rose early. After some deliberation, she selected black trousers and a dotted crepe blouse. She wanted to look nice for the meeting that Adam had proposed, with himself and the lawyer. Pippa, who was accustomed to Mother in her cycling shorts, skirt and tee-shirt, was impressed, and the atmosphere became festive.

As always, I had errands. Because of the weather, I decided to wear my warm coat, the one I had been putting on Mother. Her keys and tissues were in the pockets, so I left them on her dressing table, intending to put them back later. I was in the hallway when she exploded. 'Stop, stop! You're not going anywhere!'

Terrified, Pippa came running.

My mother, her face in a fury, was rifling my pockets. 'Jerk! Jerk! Where are they?' I knew that she meant her keys. With a sick feeling in my stomach, I told her.

'That's not good enough,' she bellowed, then clumped off towards the bedroom.

Pippa looked with big eyes. Not having seen my mother in a full-blown temper before, she was as upset as I was. 'What happened? What's the matter?'

'Nothing,' I said. But Pippa was alarmed. Everything had

been sunshine and light. She just couldn't understand the cata-
clysmic alteration in Mother's behaviour.

It would not have been appropriate to explain how sudden
rages had always been the currency of Mother's personality. I
could not tell her how Patty and I had grown around these rages
like carrots about a stone, how they had shaped us fundamental-
ly. We had lived walking on eggshells, filled with anxiety, pes-
simism, insecurity; trying always to please, but never succeeding.
Understanding that we were second-rate to deserve such relentless censure.

My father coped by switching off. 'Your mother is on her
high horse again,' he would say gloomily, then vanish into his
newspaper or his work or the bedroom. He kept different hours.
Early to bed and early to rise, he escaped to his uptown office,
seven days a week during the winter. And while Mother stayed
up all night, reading in the living room, Daddy went to sleep.

Like my father, I hid when Mother raged. Patty took the
opposite path. She learned to rage back.

⁂

Adam was the first to arrive. As always, his presence was a tonic
to Mother. His sense of fun was infectious, and she couldn't help
but be cheery when he was there. His handsome face interested
her: she reached up and touched it. 'What's this?' she inquired.
'And this? And this?'

Sheepishly, Adam muttered something about skin cancer. I
marvelled. In spite of cataracts, my mother's powers of observa-
tion were remarkable. She could see exactly what she wanted to
see.

By the time the lawyer arrived, she was vivacious. Bit by bit,
Adam steered the conversation towards homecare. With much
easy banter, he convinced my mother that she needed helpers,
four hours daily, seven days a week. 'Well, we're all agreed then,'

Adam drawled, and everybody nodded, no one with more enthusiasm than Mother.

'What my mother would like best,' I added, 'is an agency
that employed only men, all of them ten years her junior.'

She endorsed this heartily. 'That's the way I am. I don't like
women around me.' 'Ivy,' she said, 'shall we divulge the family
secret?'

The one that she had in mind turned out to be my father's
quiet presence. After all, his ashes were there with us, tucked up
in their shopping bag next to the fireplace. For some reason, we
found this hilarious. 'When you die, Mrs. Eberhart,' I added, 'I'll
burn you in that very fireplace. A do-it-yourself cremation. It'll
save a lot of money.' Again everyone laughed. It was turning into
a jolly, surreal party.

That night, I listened on the telephone to the litany of my
husband's woes in Dublin. His own mother was despondent, and
her carer was threatening to quit. Andrew was misbehaving theatrically, having decided that if only he were bold enough, I
would have to return.

'This is some miserable patch in our lives,' I whispered. For
although Mother had agreed to the principle of homecare (with
Adam and the lawyer as witnesses), this was not the same as
implementing it.

I couldn't get to sleep. The audio books, which usually lulled
me into stupor, weren't working. I listened to Virginia Woolf's
Mrs. Dalloway with growing irritation. How unbelievably silly the
heroine was. Who cared about her stupid party anyway?

❧

We were talking with Pippa in the kitchen, reliving the previous
afternoon with the gentlemen callers. 'My Mom's going to have
a little help every day,' I said, 'Mondays to Sundays.'

'I agreed to that?' my mother said, with some amazement.

'You certainly did. Don't you think that's a good idea, Pippa?'
Pippa nodded. 'It sure is, Mrs. Eberhart.'

⚜

The weather grew colder. Black ice glazed the footpath. I cancelled our plan to go to a movie. Mother was livid, so the next day, we went anyway. The doorman looked in disbelief when we appeared in the lobby. 'You be careful out there, Mrs. Eberhart,' he said. 'It's nothing but ice.' We headed into the wind, picking our way over the slithery surfaces. The cold seared our cheeks, but with my arms bracing her, Mother advanced pluckily.

I had chosen *October Sky* because it was playing nearby, and there were no stairs. I enjoyed it as much as any movie I have ever seen, because I was glad to be at the movies again. Even for me, the ordinary had taken on a special timbre.

⚜

The day of the cardiologist's appointment did not begin well. The blood man telephoned at 7 am, which meant getting Mother up early. Then, without warning, Pippa did not show up at all. Mother's humour deteriorated. Angered by her bank statements, she stumped about swearing. 'Jesus, you're a jackass. All you do is confuse me.' She was in no hurry to get ready either. When I told her that we would be taking a taxi, she blew up. 'Shut up!' she screamed. 'Don't waste my life.'

I got her into the shower and left her there. I sat in the living room breathing slowly, trying to calm down. After fifteen minutes, I went back to the shower to dry her feet.

The cardiologist's Park Avenue waiting room was well-appointed with dark leather. There was a good chair for my mother, with stout arms that she could use to lever herself. The chair delighted me. Anything that made life easier filled me with happiness. As we waited, I counted my blessings. We had made it safely and on time. The weather had improved. The doorman had hailed the taxi and helped Mother in. Getting her out had passed smoothly. And, best of all, her temper had improved. As always in a new place, she was alert and curious. Tilting her head, she drank in every detail, studied each new face.

When her various tests were completed, we were ushered into an inner sanctum. The middle-aged cardiologist sat behind a big, uncluttered desk. He was immaculately dressed and tanned. The photograph of a glamorous young woman and tot dominated the gleaming desk.

'Oh,' my mother said, 'is that your daughter and grandson?'

The cardiologist frowned. 'Actually,' he said icily, 'it's my wife.' When he spoke, he directed his words at me. 'Your mother's heart is in excellent condition.' This surprising news both startled and disarmed me. What it was, as I realised later, was a ploy to be rid of us.

What I should have said was, 'Oh, that's wonderful news. So why is she taking digitalis, verapamil and warfarin?' Or, I might have said, 'Then why are we here at all? Why are you taking her money?' Instead, I probed what he meant by 'excellent', which grudgingly elicited the information that she had a leaky mitral valve.

What was a leaky mitral valve?

He answered in monosyllables. And so it went on. This cardiologist was not going to waste the pearls of his wisdom on us. He was certainly not going to mention her atrial fibrillation, or what he was doing about it. I was sorry that I had not drawn up a long and tedious list of questions, if only to prolong his agony. And I couldn't even ask if his drugs were confusing Mother, not with her sitting there.

'I'm not going there again,' Mother said spiritedly.

'A stinker,' I agreed. The doctor's lack of interest in his patient had been stunning. A few days later I passed the same heartless cardiologist in the street, a pair of designer shopping bags complementing his cashmere-clothed arms.

❧

Getting home from the cardiologist's proved more stressful than getting there. Mother wanted to eat out and to walk home but the practicalities made me uneasy. Home was fifteen blocks away, three times farther than she had walked so far, and there were no invalid toilets en route. But Mother was adamant. At her insistence, we crept an extra block away to the restaurant she wanted, adding to the marathon.

We did not have a good time. In the restaurant, she ordered copiously, then quarrelled with the waiter when he didn't bring what she expected. When we left, I was carrying an oozing doggy bag full of the food that she hadn't eaten, which made it harder for me to support her as we inched along. Again I proposed a cab. Surely we had done enough for one day?

'No!' So we made our way through the grey streets, my mother's pace growing slower and slower, as exertion took its toll.

Home was in sight. Another block or so, and the challenge of the stairs would be in front of us. Mother ground to a halt outside Dunkin' Donuts. 'I want a donut,' she said. There was a heavy, plate glass door to struggle through, and a long queue inside.

'I'll get you some later,' I said.

'A donut.'

'We're going home.'

'Oh no we are not!' Then, in the middle of Second Avenue, my mother began to yell. Before my eyes, her gaping mouth grew larger and larger, like the mouth in Edvard Munch's painting *The Scream*. Mother stood, howling into the sky, an ancient toddler throwing a tantrum over a donut.

The next morning, when I asked Pippa where she had been the day before, she waxed eloquent about some family gathering. 'That's nice,' I said, but I worried about what would happen if she didn't turn up when my mother needed her.

In Central Park, I lost myself in an avenue of trees, breathing in their clean air. Afterwards I went to lunch at a French restaurant on Lexington Avenue. There too, I had the best of times, observing the minutiae of this slice of Paris in New York. As I ate, I thought about Patty. Had she ever studied the proprietor, a fashionably dressed Madame, and the way her laser eyes kept her minions in order? Certainly, the Alliance Française wasn't a block away, and Patty must have passed frequently. But try as I might, I couldn't see my sister enjoying lunch on her own in a smart little restaurant. Why not? I thought. Why hadn't she learned to loosen up, to like herself, to enjoy life to its fullest?

Now that Mother could negotiate the streets with help, she reestablished her habits of eating out and filling the fridge with doggy bags. At a chintzy Chinese place on Second Avenue, she was greeted like a long-lost friend. Where had she been? What had happened? My mother launched into a rollicking account of her accident, and the eleventh-hour rescue by her handsome cousin.

Later, while we were eating, she found a fat minnow dead among her snow peas. She left it on a side plate, but it did not deter her from bringing home the remains of the meal for later consumption.

I went to the nurse practitioner's office. Could the drugs be contributing to my mother's confusion? Nadia shook her head. 'A more likely explanation', she said, 'is shock: at your sister's death; or her own accident; or both.'

I asked about carers.

'Good carers,' Nadia said, 'are exceedingly hard to find.'

On the apartment floor, I had found a yellowing leaflet about Health Outreach, an organisation for the elderly. Was it possible that they could advise me about carers? I walked in off the street to be whisked into the director's office. As I told him about Mother, he offered a wealth of information and advice. I had come to the right place, and my mother, by leaving the leaflet on the floor, had directed me to it.

Health Outreach recommended four homecare agencies, so I phoned them. All had pleasant, helpful spokeswomen, and I made a couple of appointments. Then I telephoned Pippa's agency. There I was dealt with differently. Why did I need to see someone? I was asked. 'Can't the matter be dealt with over the telephone?'

Over the telephone? Arranging care for an elderly parent which was going to cost thousands of dollars?

At the lawyer's, I talked about Mother. How would she manage after my return to Ireland? What if she locked herself out? Would she venture too far, and be unable to get home? Would she leave on the gas, or the ancient toasting oven? And how would she cope with her medications, especially the warfarin, which was dangerous when taken incorrectly?

The lawyer listened. He seemed to realise that it helped me to talk my way through a situation that appeared impossible.

In the lift, a neighbour mentioned an article in *The New York*

Times about carers. Soon a copy appeared under our door, a worst-case litany of the abuses inflicted by the unscrupulous upon the confused. It fed my anxieties. I hid it from my mother.

Too early for my appointment with the homecare agency, I slipped into *Bon Pain* for coffee. Good bread, in French, but everything seemed bad to me. The city outside was formidable and studded with pitfalls. The fact was, I didn't want to hire carers. Just like Mother, I didn't want strangers in our private lives. I wanted my mother to be independent, the woman she was, the devil I knew. I wanted her to be well, and my sister to live again. I wanted to turn back the clock.

Among the tangle of Mother's papers, I had found a living will, signed seven years earlier. On it, she had added in her elegant writing,

Patricia Eberhart – my daughter. Ivy Bannister – my daughter. Each or both can push me over the side when it is time. H. Eberhart.

<u>*Do it*</u>. *Push me over the side immediately. Don't be afraid or feel guilty. Do it.*

The memory of these bald words increased my gloom.

The manager of the homecare agency proved to be its owner. She began by saying that they tried to match carer to client for a long-term relationship. When I mentioned that my mother got along with Pippa, Ann Kaplan made a suggestion: that Pippa change agencies, that Pippa work for her. This lateral thinking impressed me; and I promised to try it.

I came away with a grain of hope. I knew there would be difficulties – matching carers with my mother would never be straightforward – but for the first time it seemed possible that things might work out.

At home Mother was consumed by complaints. This time, her targets were Pippa and the physical therapist. They were 'colluding and conspiring', 'interested only in themselves'.

'But you like them.'

'I don't like anybody!'

Apparently, the two women, and the nurse who supervised them, had all been in the apartment at once. According to my mother, the women were letting the nurse know what a good job they had been doing, and, in my mother's eyes, this was outrageous.

'Come on, Mom. They're two nice women. Why shouldn't they be proud of a job well done?'

I should have thought before opening my mouth.

Mother exploded. 'Jesus you make me boil! Dissent between mother and daughter is so fucking ugly. I can't even look at you, I'm so ashamed of you. You harm me, harm me. You have a mean streak, but I don't.'

That same night, improbably, we had a lovely meal together at a cheap and cheerful Mexican restaurant of my choosing. Each table was painted with cactuses, sombreros, fruits and flowers. Bright sculptures vied with one another for attention, and mirrors were everywhere, throwing the colours back and forth. It was not a place to be sad in, and for once, my mother relaxed into calm conversation. It was, I thought, the way things should have been all the time.

After putting Mother to bed, I made a phone call. By chance, the unusual name in the Manhattan telephone directory had caught my eye. I thought it might belong to a woman I had grown up with during my Connecticut summers, and it did. We had an amusing conversation, and made arrangements to meet.

In the morning, Mother's eyes were wide open. 'The physical therapist is a monster,' she said.

'She certainly is,' I agreed, wary of provoking another outburst. The thought of mentioning the woman from Connecticut crossed my mind, but I rejected it at once. Mother always had been unenthusiastic about our having friends. While clearing the apartment floor, I had found a list of instructions that she had left for us, when she set off around the world. We were to have *no* visitors, she had written. There wouldn't be time.

I was eleven, Patty was fifteen, and Mother was away for three-and-a-half months.

Pippa decided against changing agencies. I was sorry but philosophical. If the fates were against her becoming my mother's carer, there were probably good reasons.

I stood in front of the bright red doors on 70th Street, that had once housed The Lenox School. Now the private girls' school was gone, and the building had become The New York School of Interior Design. 'Just walk right in,' Patty had told me. 'It's really interesting.' I crossed the familiar threshold for the first time in thirty years. Sailing past the security guard, I made my way to the stairs that led into my past. I knew them intimately, each baluster and handrail, from thousands of climbings. Up and up I went, the air jostling with old names and old faces. I prowled the building, relived scene after scene from my childhood.

In the riot of images that tumbled and churned, I looked for my sister, but nothing fixed on her. I could find only glimmerings. I saw her shadow on the wall where the Senior Staircase used to be. I heard the echo of her voice singing *In Dulce Jubilo*, as Christmas assembly commenced, but the substance of Patty eluded me. Was it because we had been distant from one another even then? Or was it simply that she had been so unhappy at school?

A condolence letter from an old classmate had included the sentence 'I do not think anyone in our class really appreciated her', and it was true. For her twelve years at The Lenox School, Patty was an outsider. Night after night, she wept, aching for popularity, but school had set a pattern. Patty never believed she was valued by the people she most wanted to impress. 'I can't compete,' she would say of college friends, who became bankers, legislators, society icons. 'I'm a failure.'

When I returned to the apartment, Mother scowled. 'I have serious work to do,' she said, 'but I can't do it, because you depress me so much.'

❦

It was 7 pm. At Mother's request, I helped her to bed. 'I hope you're going to bed now too,' she said. I nodded.

I didn't go to bed. Instead, I let myself out. In the brisk city night, I eased into a racing walk. Suddenly crazed for sugar, I bought some cherry liquorice which I wolfed, striding farther and farther downtown.

❦

Sunday morning. In St. Vincent Ferrer, I prayed again for my sister's soul. Then I walked the few blocks to her building, and wrapped my fingers around the door knob. Whatever about The Lenox School, I felt Patty's presence, there outside her apartment.

❦

In the kitchen, Mother was fussing with papers. In the living room, I was polishing the floorboards with a duster. A bubble passed through my body. My underwear was soaked in blood. I stared at the scarlet flood with amazement. Since I had arrived in America, I had forgotten periods. Now, after two months, my body was reasserting itself. In September, I had put Patty's tampons in the kitchen bureau. I hobbled out, clenching my thighs.

'This is wrong,' my mother said, peering at a statement. 'Can you explain it to me?'

'In a minute,' I said, grabbing the tampons.

'Now,' Mother yelled. 'You're supposed to be helping me!'

'In one bloody minute,' I yelled back.

'I never dreamed that you would behave so badly,' my mother roared. 'You're ruining everything.'

Indeed I would be ruining everything unless I used a tampon.

⁂

Mother was making lunch. She selected a doggy bag from the assortment in the fridge. Inside were the remains of her chintzy Chinese dinner, now five days old; the dinner that had included the cooked minnow. She opened the carton, left it on the counter. When the contents had warmed to room temperature, she ate them out of the carton. Mother did not believe in microwaves or making dishes.

Her stomach, I figured, was cast iron.

⁂

15 March: the final week of Mother's rehab. Rehabilitation is a powerful word. It promises restoration, or return, to one's proper condition. But how could my mother ever be what she was before Patty died?

And even though Mother's physical condition had improved, there was a distance to go. She could not get out without assistance, or carry groceries; she couldn't tend to her feet or pick up things; she could not manage more than a few stairs, or get in and out of the bath.

In the orthopaedic surgeon's office, we waited amongst the walking wounded: men, women and children, bandaged and braced. After an hour and a half, the squat, muscular surgeon paused in the corridor, a few feet away. There, he became absorbed in conversation with another medic about golf, which meandered around its eighteen holes.

Two hours after her appointed time, my mother gained the surgeon's attention. His interest in her was workmanlike. Her bones were knitting well with the implant, he said. But she needed to persist: with both the exercises and the physical therapy.

A new physical therapist would have to be found.

❦

'Absolutely not,' Mother said. She was refusing to wear a personal alarm. In spite of the 55 miserable hours that she had spent on her bathroom floor, she saw no reason to have one. Adam suggested a mobile phone instead. The alarm was simple: a panic button on a necklace. The mobile phone was complicated. There were buttons to press before making a call, recharging to remember, and phone numbers to key in. And the phone had to be carried rather than worn. It did not seem the best choice for an elderly woman and a technophobe who was having difficulties with numbers.

Why was Mother so stubborn? Why couldn't she see that a personal alarm would preserve the independence that she craved?

❦

Happily, I discovered a physiotherapy centre only two blocks from Mother's apartment. For years, I had passed without registering it. And there were plenty of male therapists on its books.

❦

Striding along the river, a blue-green dazzle caught my eye: a duck on the swell. This surprising manifestation of Manhattan wildlife delighted me, as it darted and bobbed. In the distance, planes were taking off from La Guardia airport, each a soaring diagonal in the sky. As soon as one was up, another followed, then another, not a minute between them.

A few miles away, another airport, JFK. Across the Hudson, a third, Newark. Thousands of planes more, launched into the sky around the clock, all serving one metropolis, too many to count.

So why that plane in particular? Why Swissair 111?

In Starbuck's, I was nursing a latte. Next to me, a woman in red pored over paperwork. Her sausage-tight skirt rose high over slim thighs; her bare arms gleamed. Her dramatic jet black hair was not quite clean. On a chair, an exquisite leather briefcase. When a colleague came in, she purred, a husky, ambitious drawl. A King Charles spaniel waited outside the window. Hers. Its name was Sweetie Pie.

I closed my eyes. I remembered myself as a teenager, open to every influence around me. Had I stayed in New York, had I not moved to Dublin, would I have become this woman in red?

Starbuck's was a joy. I liked being there, alone with my coffee and my diary. But I liked the buzz too, the opportunity to look around, the pleasures of privacy in a public place. Patty had introduced me to Starbuck's years before. As so often with her, the occasion had been more stressful than fun. She had fussed over everything: the correct coffee; the change; the best seats. She wasn't able to take things as they came, even the little things.

A couple of years later, I had suggested a second visit to Starbuck's. Patty had a flight, and I was returning to Ireland. We wouldn't be seeing each other for a while. Starbuck's would be a good place for goodbyes, I thought. We could sit, relax and talk.

'No,' Patty insisted. 'I've too much to do. Come to my apartment, and I'll make coffee there.'

As the door to her apartment creaked open, I could see she was in a dark mood. She was wearing only an ancient slip, inky crêpe de chine. On the back of her shoulder, I noticed a mole: raised, dark and ragged.

'You should have that checked out,' I said.

'Should I?' she barked. 'Do you think it would be enough?

Enough to do for me, if I left it alone?' She slammed around the kitchen, making coffee. She made it seem an incredible effort. I sat in the living room, trying to occupy as little space as possible. Conversation was out of the question. Whatever I said provoked gall. I thought of Starbuck's, and its cheerful, bantering service. At last, the two cups were ready. Scowling, she brought them out, large, green Italian cups; and she sat on the spare bed that doubled as a sofa. As she raised her cup to her lips, its handle separated cleanly. The cup tumbled to the floor, spraying its scalding contents over her scantily clad body, soaking the bed and her Persian carpet.

Patty howled with pain and rage.

I tried to help, but she snatched the blotting towels from my hands, swearing at my incompetence. Didn't I even know how to clean something up? She tore the sofa bed apart. The coffee had soaked the padded quilt, the sheets, the mattress protector. The brown stains seemed to grow as we looked. 'What's the matter with me?' she roared. 'Why can't I even drink a cup of coffee like anyone else?' She hauled the bedding off to the tiny sink, refusing assistance, deaf to my calming words. 'When the New Year comes,' she hissed, 'I am going up to the roof of Mother's building, and I am going to jump.'

I felt helpless and appalled. She could take nothing from me. She could give me nothing but rancour. We couldn't communicate, not about the things that mattered. We could not offer one another the friendship that each of us craved. And, that day, as on so many others, I couldn't wait to get away from her.

Mother woke up crying, the tears chasing one another down the tracks of her wrinkles. To stop them, I invented questions about where the television cable (which I had ordered against her wish-

es) should be installed. Soon, she was out in the living room issu-
ing instructions.

'OK,' I said. 'When Pippa comes, we'll move this table here
and that bureau there.'

'No. I'll do it with you now.'

'Don't be silly. All you need is to put too much pressure on
that healing bone —'

'I'll do it. It's all in the arms.'

Delicately, my mother and I shifted the furniture.

Pippa had good news. She was leaving her agency to work in
a public school library. I was delighted: she deserved better than
a starvation wage from a third-rate agency. I was also relieved
that I hadn't been counting on her.

The lawyer's office. He had a surprise, he said. He handed me
Patty's cheque-book. 'Look in the back.' I poked in the hidden
pocket. Inside I saw the green glow of folded bills. I dropped the
cheque-book into my handbag.

The lawyer took me to lunch. We walked westwards, across
the tip of Manhattan towards the Hudson River, through crowded streets and massed buildings. As far as I was concerned, the
lawyer was a hero. I would have been lost without his hours of
tedious application, and his advice. Once again, I had my mother to thank. It was she, after all, who had selected him.

Back in the apartment, I looked at Patty's cheque-book with
a terrible sense of trespass. What right had I to her cheque-book?
I turned to the hidden pocket and exposed her cache. There were
five twenty dollar bills, each folded as precisely as origami. Before
me, Patty's long fingers danced, perfecting the crisp folds of her
hard-earned money.

When talking with the owner-manager of the homecare agency,

I had described Mother as talkative and well read: 'full of ideas'. I had also mentioned her volatile temperament and hostility to visitors. In response, Ann Kaplan had proposed Clare as a carer, and I could see why. For, like my mother, Clare was bright, talkative and full of ideas. But my heart sank as I watched them together. Clare treated Mother like an equal, and Mother didn't like this, not one little bit. For although my mother needed to talk, she wasn't interested in listening. It was her own views that fascinated her, and nobody else's. She didn't want to be challenged; she didn't want to engage. Endorsement soothed her; debate unsettled her. She needed a companion who would listen only, and be impressed. I knew this all too well from my own experience. Unfortunately, it hadn't occurred to me to define it for Ann.

After Clare left, Mother let loose. She loathed Clare. Clare was impossible because she was fat, because she was black, because she wore a baseball cap, because she talked about her own dead relatives. What did my mother want to know about *her* dead relatives, for God's sake? On and on she raved, annihilating the woman from every angle. Eventually my mother clapped a hand over her heart. 'I was fibrillating every minute she was here,' she said theatrically. 'The fact is, I don't need any help. I can look after myself.'

That night I explained to my husband what had happened with Clare. Frank's loneliness and stress floated across the Atlantic, the mirror of my own. After more than two months, I was no closer to coming home.

Mother's oedema continued to alarm. Her ankles and feet were balloons of water; touching them left indentations. Cortisone cream kept the sores at bay, but diuretics had little effect. The swelling seemed to get worse daily. I encouraged her to put her feet up, only not too far, on account of her hip. Additionally, I worked on her soles with moisturiser, and trimmed her toenails.

⁂

Another Monday. I wrote out my weekly list of what needed to be done. All the significant entries had to do with the question of Mother's home care. I saw to it that she breakfasted, took her meds, did her exercises. The blood man came and went. Then, as if Pippa had arrived, I headed for the door, telling Mother I would be back at the usual time.

'I'm going back to bed,' she said resentfully.

Adam accompanied me to the homecare agency. With nothing but praise for Clare, I explained how I had omitted an important element in my mother's portrait. Once again, Ann Kaplan was full of ideas, among them the suggestion that we refer to the carers as housekeepers. This, she thought, might make a psychological difference to my mother. She had another woman in mind.

I crossed my fingers.

⁂

There were tickets to *Madame Butterfly*, the last opera in Mother's subscription. In spite of the distance, the crowds and the stairs, I was determined to get her there. The third ticket, I thought, should go to some able-bodied person who could help, someone like Sylvia, my recently discovered friend from Connecticut.

'Sylvia?' my mother said contemptuously. 'What can she do for me?'

'She can help get you there.'

'No, I don't want Sylvia.'

'Who do you suggest then?'

She thought for a long time, then finally said, 'Mrs. Quincy.'

Mrs. Quincy was a perfectly nice woman who lived down the hall. She was also in her nineties. 'Mom,' I said, 'if you got dizzy, who would you rather have with you, Sylvia or Mrs. Quincy?' Sourly, Mother launched into a diatribe, but in the end, we invited Sylvia. Sylvia arrived in a glamorous whirl of red coat, dress suit and ash blonde hair. Full of fun and sparkle, she had Mother laughing as we eased her into a taxi. At the opera house, we flanked Mother, protecting her from jostles, supporting her on the stairs.

I watched *Madame Butterfly* with the ghost of my sister. 'It's too sad,' I heard her whisper in my ear. Patty had believed in happily-ever-after, in the nobility of men and romance. *Madame Butterfly* had depressed her with its alternative scenario.

❧

Mother came back from the refuse room with an armful of newspapers, a cigar box and catalogues. She deposited them on a chair that I had cleared for sitting. When she went to bed, I returned everything to the refuse room. The next day, she referred to the cigar box, but fortunately didn't look for it.

Even though Adam was coming to take Mother out to lunch, she was cranky. The idea of housekeepers remained anathema. 'The lawyer and I can handle everything,' she insisted.

'Certainly he will keep your finances straight, but he can't help with personal care.'

'Yes he can. And don't confuse me. You're very bad for me. You need to listen. And you should never repeat what I say to anyone else. It shows a lack of sophistication.' Then Mother turned on the TV. Soon she was laughing uproariously. 'Look, Ivy, look! Do you see how fat that woman is? Does she know what she looks like?'

❧

Mother was telling me how I should bring up Andrew. 'Smack him around,' she insisted. 'He'll learn faster that way.'

'Just look what it did for Patty and me,' I said.

❧

The new housekeeper proved to be a trim woman of my own age. She came into the apartment uncertainly. At once she mentioned her most recent client, and how she missed him. My mother prattled away, veering from subject to subject. Soon, she was talking about my father. 'Of course I made plans for my husband to go into a nursing home,' she said, 'but I didn't implement them for a year.'

Made plans for Daddy to go to a nursing home, which she didn't implement for a year? The long years of my father's decline rushed into my head, and my mother's refusal to acknowledge it. 'What are you going to do about Daddy?' I had asked, suggesting an Alzheimer's centre in Dublin.

'There's nothing the matter with the man! He could act better if he tried. He's just perverse. He does what he does to annoy me.'

Matters came to a head when she became executor of a will. Exhausted by caring for Daddy, she hoped that a week's respite would help her manage her executor's duties, so she booked him into a place in Connecticut. But within an hour, my eighty-year-old father had punched a female attendant, scaled a six-foot wall, and made an escape. Pursued and restrained, he was carted off to a psychiatric hospital. Mother was informed that she shouldn't even try to bring him home again. And so, under duress, a nursing home was found, where he was bound into a wheelchair with a pelvic restraint. It was a heartbreaking crisis, made worse by Mother's refusal to address his illness. Now, it seemed, she had invented a new scenario.

Mother barrelled on to a new topic, then another. Finally, she turned to the listening housekeeper and said, 'Are you going to work for me?' My mother, who formed lightning judgments, had taken to Daphne.

❧

An elderly cousin died. Adam proposed that Dorathea's funeral become a celebration, not only of her own ninety-seven years, but of Patty's life as well. Persuaded by Adam, my mother agreed. Just what, I wondered, would Patty have thought? My sister, who had planned her estate considering others, had left no instructions as to what we should do for her.

❧

Mother could not find her favourite book, *The Lives of the Cell*. 'You threw it out!' she bellowed. I shrivelled with fear. My eyes scanned the shelves, locking on to the slender spine. I handed the book to her, but she didn't calm down. She sent me out to acquire two more copies. 'And don't come back without them,' she yelled.

'Read *Lives of the Cell*,' I heard her say, as I walked down Third Avenue. 'If you read it, you'll understand *everything*.'

In my head, I saw Patty frown. 'I've read it, Mommy, and I don't understand *anything*,' she whined, in a voice that mixed aggression with self-pity.

❧

Mother had a question: 'Owing to your sister's close relationship with the accountant, should he be informed of the memorial service?'

Patty's close relationship to the accountant? She had seen him once a year professionally. I had been there, the first time. Patty had come out of his office in a temper. What had he done for her, she had demanded, that she couldn't do for herself?

Then Mother suggested Patty's supervisor from the airline. But Patty had insistently kept her personal life separate from her employer.

'Why not some of Patty's friends?' I said, mentioning a few people that Patty liked.

'No. Absolutely not.'

I was beginning to feel uneasy. I had mentioned the service in passing to Sylvia, who had made a note of it. I hadn't asked her to come, but she was apparently thinking about it. When I told Mother, she exploded. 'How dare you?' she thundered. 'Don't you know anything? Sylvia? Why, she'll think it's a party! My God, to think that my daughter is such a jackass. I'd be so embarrassed if Sylvia showed up.'

It was eight o'clock on a Monday morning, and already I felt wretched.

In the evening I called Sylvia. 'By the way,' I said, 'about that memorial service: I've just realised that it's for family only.'

My friend was not fooled. 'I'm reading between the lines, here,' she said good-humouredly. 'Even when we were kids, your mother made it obvious that she didn't want me around. Not me, or any of the others.'

Another woman arrived from the homecare agency, a motherly figure who was willing to work weekends. 'Everybody likes me,' Marianne said, and miraculously, Mother did too. When Marianne left, I discovered that she had polished the drain in the kitchen sink. That gleaming O struck me as a favourable omen. On the strength of it, I booked my return to Dublin. In ten days, home! The idea filled me with joy, but my emotions were turbulent. Even with highs ahead, the lows remained consuming.

The first day of April. I was in Central Park, trying to shake off my worries. With a bellow, a helicopter buzzed overhead, so low that my scarf flapped. The initials, NYPD, were emblazoned on the machine's sides: New York Police Department. What was up? What murderer or gang of thieves was lurking in the nearby bushes? An army of vehicles loomed ahead: police cars, ambulances, a clatter of outside broadcasting units. A crowd had gathered too, behind barriers which were being patrolled by mounted policemen, whose horses' hooves sparked off the pavement. I finally spotted the culprit: a shaggy, brownish animal in a grassy hollow, who danced away from a posse of captors bearing truncheons, nooses and rifles.

At the barriers, a pudgy cop held court. 'It's a coyote,' he smirked, a hand on his bulging holster. A coyote? I looked again at the dancing animal, balletic as it evaded its tormentors, unable to catch it for all their implements of destruction. It was a surreal sight: hunters and hunted against a backdrop of skyscrapers.

I was for the coyote.

A gentle tap came on my shoulder. A Japanese man bowed. 'Do you know, please, what is happening, please?'

'It's a coyote,' I said, but he didn't understand. So I spelled it out. He listened diligently, repeating each letter.

Suddenly, he got it. 'Ah. Kai-yo-tay!' he cried, turning to his comrades, a compact group of Oriental gentlemen, who bandied the word back and forth like the name of some rare, delectable wine.

Now, alas, my coyote was slowing down, whether from exhaustion, or from tranquillising darts, I didn't know. Nooses tightened round its neck, and it toppled over. Grabbing hands strapped it on to a stretcher, then whisked it away in a cacophony of sirens and screeching tires.

The scene changed swiftly. The chopper rattled off, the crowd dispersed, the horses trotted away, the emergency forces folded themselves into their vehicles and vanished. Only I lingered on with the squirrels, the smell of the hot dogs being sold on nearby Bethesda Terrace, and my head full of the coyote, having forgotten for a while the anxiety that had brought me into Central Park in the first place.

❧

I met Emma, Adam's wife, outside the church in the East Eighties. Emma had her dog, a delightful, loping animal with a mind of its own. We were to meet the minister to discuss the service for Dorathea and Patty. But first, Emma took me (and the dog) on a tour of the church, a comfortable New World building, built in the style of the French Renaissance.

In the busy parish offices, my eye was drawn at once to a picture of Christ reading *The Village Voice*. The minister himself proved equally irresistible, with a pate that gleamed like mother-of-pearl. He, too, had his dog, a fluffy white poodle, and after various doggy exchanges, we settled down to business.

This was new territory, the minister observed: the celebration of two lives at once. It hadn't been done before, not so far as he knew, not with the two subjects dying at different times and places, and certainly not with one coffin in the church. Whoever spoke in memory of the dead, he said, should be sure to mention both women in their remarks.

He was a lovely man. Patty would have liked him.

As the Fates would have it, Emma's mother died the following day. And so Emma found herself caught up in the death of her own mother, arranging two funerals (for three people) in two states at the same time. And somehow I had imagined that the wings of the angel of death fluttered only in my vicinity.

Writing the eulogy for Patty, I disappeared into the past, in search of the best memories. At night, I dreamed she had adopted the baby that she yearned for. In that dream, Patty was happier than she had ever been. 'But oh,' she said, as she cuddled and loved her new baby, 'what a lot of work!'

❧

Monday, 5 April. In the pulpit, I waited for the congregation to settle. I smiled for all I was worth, smiled so I wouldn't cry. Then Patty's fine qualities filled the air: her generosity, her ability to touch lives, her oneness with children and with Christmas. Like my father, I said, she was passionate about the stock market. Like my mother, she was passionate about travel. And like herself, she was passionate about art. I touched upon the physical resemblance between her and my youngest son, and how, when I looked at him, I would always see her. Then I concluded with a poem I had written for her.

My mother was pleased, which is what mattered the most.

With Dorathea's remains, we drove out to Woodlawn Cemetery, a vast necropolis, an incomprehensible density of bones. Dorathea's coffin was lowered into their midst. As it descended, I travelled back to her apartment, and the last time I had seen her, a couple of years earlier. We were sitting together in her front room. 'I've had a wonderful life,' she told me; 'a wonderful life.' She said it again and again, her face radiant with the satisfying words.

It was not, I thought, how Patty would have summed up her own time on earth.

We returned to Manhattan to lunch in Adam's and Emma's home, where I met many of the new generation for the first time. The lively young faces reassured me. There were so many of them! The youngest was just walking; the eldest was in her twenties. Death had made me see life in a new way, and the young people thrilled me.

That night, I thought about Patty. As an adult, I had been frustrated by the difficulties between us. Mostly, we could not have a good time together. Like Mother, Patty would criticise my clothes, my shoes, my hair, what I was ordering for dinner, the way I walked, the way I moved my hands when speaking. We could not have a conversation without her getting irritable, and sometimes furious, because she disagreed so strongly with what I thought, felt and said.

But unlike Mother, Patty was even harder on herself. 'I'm just a dum-dum,' she would say, then launch into a spiral of negatives: life had passed her by; she had missed out on marriage and kids; she didn't have enough money; her job wasn't good enough nor was her apartment; her background was lacking; her education lousy. Being with her was like being with semtex, knowing that explosion was coming. Yet all the good things I had said about Patty at the memorial service were true too. Why had they not ruled her personality, enriched her relationships? Why had she not matured into a happy, confident woman who enjoyed life?

The following morning, a pilgrimage. I walked uptown to 103rd Street and Fifth Avenue, to the Conservatory Garden, near the northern boundary of Central Park. I had been there only once before, seven, perhaps eight years earlier, with Patty. It had been a good day. Patty was happy. She had been off work for months with her broken ankle. Now, much rested, she was walking fluidly, and the stress of her working life was on hold. We had Richard with us, who delighted her with his ten-year-old's enthusiasms. And we had my father, unpredictable in his illness, but for the moment contentedly soaking up the spring sunshine.

The four of us ate our sandwiches, at peace with one another. The lilacs were in bloom, the air delicious with their scent. No one else was around – just us – in a leafy oasis, a pocket of serenity in the bustle of Manhattan.

Now I sat with ghosts. The garden hadn't changed; it was still serene. The lilacs were not yet blooming, but there was a dazzle of ivory blossom, and bright blue pansies with yellow hearts. It was we who had changed forever: Patty and my father were dead; Richard was a young man; and without my sister, I would never be the same. It was still beautiful, achingly beautiful, a superior place for remembering Patty. I could hardly bear to leave it.

I was going home to Dublin in three days, but I continued to sort and tidy. I had not yet probed under the furniture. I dropped to my knees to investigate beneath the chest of drawers in the bedroom where I slept. The few inches were packed. Out tumbled a bundle of dusty papers. They were legal documents about Uncle Pierre, Mother's brother, the man who had frozen to death in a field at the age of fifty-two. But these documents were not about his tragic death; they were dated fifteen years earlier.

I started to read. I discovered that my uncle, Pierre Jacques L'Amour, aged 35, had been declared 'an Incompetent Person' by the courts on the grounds of schizophrenia, 'mixed type'. His prognosis was deemed to be 'guarded'. He had been committed to Rockland State Hospital in upstate New York, without benefit of parole. Four years later, he was released under convalescent care. Three years after that, the courts judged him able again to manage his own affairs.

As I read through the papers, I felt a missing piece of the puzzle of my life slam into place. The fact that my uncle's schizophrenia was bad enough to warrant institutionalisation seemed an important plank of family history. Why had it been left for me to find out while rummaging under a chest of drawers? And what about Patty? Had she known? I was doubtful. She had never mentioned it to me.

My mother had gone around the world, she had said, to get away from Patty, who was a difficult teenager. The apartment had seethed with turbulence, screaming matches, blue language, angst. But for Mother to go away? Around the world for three-and-a-half months? To run away from the problem? Was it possible that confronting the situation during those teenage years would have helped? Was it possible that, as an adult, Patty would have made different choices had she known about Pierre? Might she have tackled her rages, her rooted unhappiness?

It was wrong to pretend that this piece of family history did not exist. I should have been told; we should have been told. I had asked Mother many times about her only sibling. When I was younger, she had exploded with temper, refused to answer. Later, she had blamed life, alcohol and Pierre's ex-wife for his death. She had never hinted at a medical history that might have contributed.

Two days remained before my departure. I worked obsessively on a long poem with a content that startled. Only later did I realise that the poem was a turning away from death and the past, towards life and the future.

In the evening, my husband telephoned. We chatted for twenty minutes, the easy-going ramble of intimacy. My mother listened to every word with growing displeasure. As soon as I hung up, she grumbled, 'How many times can you repeat yourself?' Then she added, 'If only you would comb your hair.'

❦

Eight in the morning, the day before my departure. 'I can't stand having you around,' Mother said. 'You are so awful. You cut out Patricia while she was alive. Now you cut me out. You've always some excuse. You're never there. You have a wall about you. It occasionally opens, but it snaps shut like the jaws of a dragon.' Then she launched into her catalogue of my multiple failures.

'Not everybody sees me the way you do,' I said softly.

'Yes they do. You're so eccentric. It harms you to be so eccentric.'

'No, Mom, your view of me is unique.'

When I was overweight, my excess poundage was one of my mother's favourite subjects. She liked to compare me with a carer in my father's nursing home. This carer, according to Mother, was unbelievably fat. However – and here my mother would pause for effect – I was even *fatter!* When I met the woman, she was indeed generously endowed. So much so that it struck me that my mother's perception was suspect.

The episode opened my eyes. For if my mother was wrong about this, might she not be wrong about other things too? And if she really did see me as bigger, well, wasn't that *her* problem? Nothing I could do would change her perspective (and when I did lose 25 pounds, it was a week before she noticed). So now, I reminded my mother of this incident as possible evidence that people did not see me as she did. 'All you do is repeat yourself,' she sniffed. 'You are a very boring person.'

⁜

The subway downtown was crowded. The large black man next to me was addressing the entire carriage on the theme of unconditional love. Loudly and cheerily, he made it clear that we should all love one another, no matter what.

Most people ignored him, but not the gentleman across the aisle who put his hands over his ears. 'Would you ever love off?' he yelled.

I was early for my appointment with the lawyer, so I wandered into St. Paul's on Broadway. There, where George Washington had prayed after his inauguration as the first President of the United States, I thought about unconditional love. Unconditional love is what one feels for babies. It is something that children themselves test as they grew older. 'Will you still love me if...?' If what? If I eat too many sweets? If I steal something? If I kill someone? Children ask these questions, because the idea of unconditional love seems suspect to them.

As we grow up, we expect things of one another: courtesy, respect, support, fidelity. Successful relationships depend on how people accommodate each other's expectations. My relationships with my mother and sister had foundered because of different expectations. I do not remember when my mother's love for me was anything but conditional. As a child, I strove to become the person that she wanted me to be. As I grew up, I discovered that I couldn't be that person. She felt cheated, and told me so constantly.

However I had my own expectations. I longed for tolerance and understanding, but Mother could not oblige. We were a disappointment to one another. Just that morning, my mother had said, 'You cut out Patricia while she was alive. Now you cut me out.' I understood what she meant. When my opinions, appearance and behaviour were met with constant disapproval, I became secretive. Silence was preferable to abandoning individuality. I couldn't please Mother – or my sister either – so I mostly gave up trying.

I perceived these things more clearly than ever. The absence of accommodation had impoverished our lives. The countless hours of misery should have been different.

Leaving St. Paul's, I headed for the lawyer's. After three months with my mother, I was drained. From experience, I knew that the last twenty-four hours would be difficult. In spite of what she said, Mother did not want me to leave. Her strong feelings about my departure would influence her behaviour. In my present state, I didn't know if I could cope.

Dignity, patience and courage, I thought – the mantra that I had used for years to calm me down when dealing with Mother.

My list for the lawyer was short. It should have taken five

minutes. Instead, I stumbled through it. The item that troubled me most was 'keys'. My mother's views on the subject remained inflexible. She was adamant that no one was to have her keys. Twice I had sat in Dublin, helpless in the face of crisis, imagining my mother dead, or in need of urgent medical attention, as she was on the second occasion. Both times, there had been no one who could gain access.

Unknown to my mother, Adam now had a key for the new lock, which he paid for, when he broke into the apartment on Christmas Day. I had had two additional key sets made. I handed one to the lawyer. But the lawyer's office was downtown. Adam's office was uptown. What if quick access was needed outside office hours, and neither key-holder could be found? The superintendent held keys to most of the apartments. Why shouldn't he have Mother's too, although she would be apoplectic? I explained this all to the lawyer. 'What should I do?' I asked.

'Give them to the superintendent.'

The lawyer took me to lunch in the South Street Seaport. And so, I found myself sitting outdoors on a sun-warmed deck with a view of the East River and the Brooklyn Bridge. Nothing could have been better. For then, as on so many previous occasions, the waters of the East River had a therapeutic effect. The brightness, the movement, the great open expanses, all dissolved my own crabbed difficulties. My eyes swept around the spectacular panorama. I ordered a glass of wine and loosened up. Soon, I was talking about the documents I had found crammed under the chest of drawers, the documents that had made me angry.

Although surprised to learn that there had been a brother, the lawyer took a temperate view. You wouldn't have expected her to mention mental illness, he argued. People of her generation didn't talk about problems. Traumas were repressed and emotions denied.

'It might have made a difference to my sister's life,' I said. 'She was terribly unhappy.'

The lawyer looked at me. 'That's not what your mother says.'

I shrugged. We moved on to other themes. I ordered another glass of wine. The lawyer, a teetotaller, looked suitably shocked by this indulgence. The old proverb drifted into my head, 'Least said, soonest mended.' There was no point in confronting my mother with the fact of Pierre's illness. As time passed, the anger would diminish, but I would never shake off my conviction that Mother should have told us.

I returned to the apartment feeling better, ready for the rapids ahead. As a prelude to packing, I set out my personal belongings. Among them were two child-sized items that I had retrieved from the clutter I had cleared from the apartment. I figured that they had been bought for Andrew, or maybe even Richard, and then forgotten. One was a hat, and the other, a cheap and cheerful fluorescent green parka. I was bringing them to Andrew, on the grounds that a 'gift' from his grandmother would be politic.

Mother was standing in the doorway, watching my every move. She wanted to know about everything: the clothes that she had seen me wearing; the cosmetics inside my cosmetics case. I gave her a potted history of every item. When I got to the child-sized hat and parka, which she didn't recognise, I said how I had found them while tidying, and that Andrew would be writing to her about them.

My mother went berserk. Capable of enormous rages, she seemed intent upon outdoing herself. She called me every name in the book, and invented some new ones. Then she grabbed the hat and stumped off, bellowing, 'You think you can take over everything! I'll have you arrested. I'll see to it that you get nothing.' Soon she was back under my nose, purple-faced, with the tiny knitted cap pulled on to the crown of her head like a pimple, the tight circumference of its band preventing further descent.

'Get out of my way,' she growled. She grabbed the fluorescent parka, yanked on one sleeve, but couldn't manage the rest. So I helped her, stretching the little garment beyond capacity. With dark language streaming, she posed in front of the mirror, an aged crone in a tiny hat and a neon-green child's parka with sleeves to the elbow and a gaping, unbridgeable front. I felt like laughing and crying at once. Instead, I apologised. But Mother was unmoveable. My behaviour was unthinkable and unforgivable. I was a monster. She wasn't going to get mixed up with anyone like me again. Even as I contemplated the ludicrousness of the scene, I knew that without the hat and parka, it would have been something else. In Mother's eyes, the really unforgivable thing was my departure.

Thus passed our last evening together. I closed my eyes and imagined the East River, the Brooklyn Bridge.

⁂

In the morning, I was cheerful and pleasant. I stripped the beds and took the sheets and towels to the laundry room in the basement. While they were in the machines, I signed in the keys with the superintendent, glancing over my shoulder, expecting Mother to materialise and annihilate me. Back in the apartment, I remade the beds. Then, for the umpteenth time, I went over my mother's medications with her.

She refused to meet my civility halfway. 'I am not going to telephone you,' she said. 'No, no, no. There is no point.'

As I dressed for the airport, I felt conscious of my every banal action. Yet again, I saw Patty, packing her hand luggage, showering, eating, washing dishes, stepping into her uniform. I saw her water the plants, switch out the lights, lock the Chubb locks. I saw her go down the stairs and step on to the street.

Each of these ordinary actions seemed precious.

I put on my own clothes: velvet trousers and a dark blouse and jacket, brightened up by one of Patty's scarves. In the hallway, Mother latched on to the scarf. 'That's mine!' she said, pulling at it. 'That's Patricia's.' I took off the scarf and gave it to her, observing that I had worn it on my journey to New York, that it had been among the effects shipped to Dublin after Patty's death.

Mother seized it, muttering ominously. 'It's all shit. If you're going to act like shit, you'll just diminish yourself more and more. I don't care about you. I care about myself. Thank God there is myself.'

I nodded, afraid to say anything.

Eventually it was time to go.

'Goodbye, Mom.'

'Don't say goodbye. Forget it.'

'I'll say goodbye anyway.' I hugged her.

'You didn't do enough,' she called after me. 'You didn't do enough for me.'

Nova Scotia

L ate August 1999. My husband, myself, and our two sons were flying north towards Halifax. The purpose of our journey was to scatter my sister Patty's ashes at the site of the plane crash. Our trip had been scheduled to coincide with the Swissair memorial, two days of events commemorating the first anniversary of Flight 111. It was late and we were tired. Earlier in the day, we had flown from Dublin to JFK, before transferring to Newark, where we had been delayed for hours. Now, on the final leg, my thoughts were with my sister, hurtling through the dark on her last flight out of New York, and into the coastal waters of Nova Scotia.

Outside the cabin window, the sky had given way to a fur of fog. Conditions hadn't been good either on the night that Patty died. There was fog, wind and drizzle. And in Patty's plane, a fire was raging in the ceiling behind the pilots' heads, a fire hot enough to melt aluminium. The pilots struggled on towards Halifax, towards the chimera of an emergency landing. 'Swissair 111, just a couple of miles,' the controller had encouraged them. 'I'll be right with you.' Then the plane suffered full electrical failure. For six more minutes, the pilots kept aloft, flying blind, their instrument screens blank, cut off from air traffic control, smoke within their hot cockpit and fog without.

I peered out the window, yearning for a light. The plane was descending; but not a flicker was to be seen. Lower and lower we dropped, faster and faster. I couldn't drag my eyes away from the window, from the murky, swirling haze, searching in vain for a point of reference. At last, something glimmered, one fuzzy flicker of light that thrilled my tense, tired body. A few seconds later, we were on the ground. We taxied, then disembarked. This is what never happened, I thought. This is where Patty never walked. This is the air she never inhaled, ecstatic to be alive.

Andrew was sound asleep; so asleep that I had to hoist him upright and propel him off the plane.

Although it was late, Halifax Airport, our fourth airport of the day, was busy. Thankfully we were collected by a volunteer, Bill, our first Canadian. We slung our luggage into his vintage white limousine. Soon, in this stately vehicle, we were sailing down a highway, the headlights carving a steamy tunnel through the fog. Accustomed to the congestion of Dublin roads, this deserted mega-highway amazed us.

A conference room in the Halifax Sheraton. I was to meet Trey Shannon of Kenyon International, a company that describes itself as Worldwide Specialists in Disaster Management. Trey proved to be a super-sized Texan. Without preliminaries, he handed me a cardboard box, pointing to its label. Patricia Wesley Eberhart, it read. 'It says the same thing on the bottom of the urn.'

To be sure, to be sure.

Trey's voice was redolent of the South. The remains of the deceased only were in the urn, he explained, and when I scattered them, a small piece of metal would fall out, a safeguard against one lot of ashes getting mucked up with another. 'Inside th' box,' he drawled, 'there's a velvet bag. Inside th' bag, there's an urn. Inside th' urn, there may or may not be some crumpled white tissue paper; 'cause where there wasn't a whole lot recovered, th' tissue keeps th' rest from shiftin'. An' inside th' urn, there's a plastic bag o' ashes, with a wire tyin' th' bag shut.'

A paper was produced for me to sign, acknowledging receipt,

together with two further documents. One was a simple message:

To Whom It May Concern.

This urn contains only the cremated

human remains Patricia Wesley Eberhart.

This was signed, J. O. 'Trey' Shannon III. Trey's moniker looked impressive, but I would have liked the word 'of' between 'remains' and my sister's name. The other document was a Death Registration, including a Medical Certificate signed by the coroner, Dr. Butt. The place of injury was described as 'Ocean'. The injury occurred by 'Aircrash into sea'. The cause of death was 'Blunt Trauma'.

The box was heavy. I opened it as soon as I got back to my room. I did not want Patty in cardboard for one moment longer. The velvet pouch slid off to reveal a shining urn.

⚜

At 9 am the following morning, we met in the lobby outside the hotel's main attraction, a casino. Already, the casino was jumping, a farrago of chinking machines, croupiers and flashing lights. We were six: the four of us; our minder Janis, who was a lively lady in jeans and an orange shirt; and a police chaplain, Canon Robert. The van headed up through Halifax, past the Citadel, down through a sprawl of fast food emporia and video shops. Soon we were again on an impressive highway: broad, straight, well-surfaced and almost empty. The day was warm. Janis passed around some sea-sickness tablets. We swallowed them dry, except for Andrew, who chewed, complaining of the taste.

As I cradled the urn, my eyes and ears latched on to details. 'Bridge freezes before road,' read a sign. 'Hubley. Timberlea. Five Island Lake,' said another. 'Upper Tantallon. Peggy's Cove. St. Margaret's Bay.' We turned off the highway. There was a notice

outside a flower shop: 'In Memory of Swissair 111.' By ten
o'clock we were at the government pier near Peggy's Cove. Great
grey waves were rolling in, smashing against the rocks in spectacular fashion. Frank looked alarmed

A boat was circling in the maelstrom. When it saw us, *So
Much to Sea* nosed in, and we scrambled swiftly on board to prevent battering against the dock. At once, the wind and waves
began to knock us around, lifting, then dropping us in a stomach-
swilling dive. The captain was young, good-looking and cheery.
'Only one rule,' he yelled, 'and that's hang on for dear life!' There
wasn't any choice. We couldn't move without holding on to
something, for the boat yanked us in every direction, firing salt
spray everywhere. Picturing Andrew's blond mop vanishing
beneath the waves, I jammed a life jacket over his head, donning
one myself. Soon, we landlubbers were green.

In the cabin, the Canon put a purple stole over his police
issue raincoat. Two or three years younger than Patty, he had car-
roty hair and glasses. The inside of his raincoat was fluorescent
orange. An unusual job, I thought, to say words over ashes he
had never known in living form. I wondered what Patty would
have thought of him, and of Janis too, strangers at her last
farewell. And what would she have made of *So Much to Sea*, once
a trawler, now used for whale-watching and ferrying remains to
Eternity?

We forged ahead towards the site of the crash. The waters
churned: blue, grey, green, turquoise, violet. I imagined a
swirling fog. I saw the immense aircraft – black as pitch – thun-
dering out of the sky in its death dive, slamming into the unfor-
giving surface, its inky sheen vanishing, its tail impacting with
the cockpit in a third of one second, pulverising the plane and its
contents.

The Captain cut the engines. We were ready. But where was

Frank? The door of the head creaked open, revealing my husband collapsed across the sink, retching. The boat was yawing.
There was no point in waiting.

The service for Burial at Sea is short. When the Canon had
done, Richard, Andrew and myself spoke our memories into the
wind. Then I took out the urn and unscrewed its top. There was
no tissue paper; the urn was full. My fingers fumbled with the tie
that bound the plastic bag. Inside, there was grey white powder,
very fine. So fine that individual grains could not be distin-
guished. Bone china. The wind was wild, flapping hair and
clothes against my face. The deck bucked underfoot. Leaning
right over the side with the urn, I began to shake. Janis hooked
her arms around me, gripping tightly. Reluctantly, the powder
slipped through the neck of the urn. The wind picked it up, frisk-
ing it about in the air, over my arms, into my face and hair, under
my fingernails. Then clumps of it dropped into the sea, white
gashes in the water which sank and dissolved. Ash and sea water.
Sea water and ash. A human being swirling in the blue, disappearing as I watched, mixing her molecules into the fabric of the
universe. As she vanished, I tasted the ash on my fingers.

There were pink and white carnations, which we dropped
into the sea one at a time. They floated, rising and falling, exquisite on the water. As the boat circled around those flowers,
Andrew and I talked about what had happened. Richard stood
aloof, solemn with his thoughts.

The journey in was calmer. We seemed to be rolling with the
waves instead of battling them. On the lower deck, Frank lay
sprawled in the open air, with a plastic bag and some paper towels. On the upper deck, Andrew was at the wheel, his ten-year-old
face in a trance, as the forty-foot boat pulled beneath his fingers.
I watched his long hair streaming, and smiled wryly. How
pleased Patty would have been by the sight of her look-alike

nephew at the wheel of *So Much to Sea*. Another new experience facilitated by his aunt. Even though she was dead, Patty was still bringing us to new places.

The contact of Frank's feet with *terra firma* brought instant relief. At once he told me, 'If you want to be buried at sea, I'm doing it by helicopter.'

We stopped at the Anglican church on the outskirts of Peggy's Cove. Perched on a little rise, the small white-timbered building was like something out of a story book, simple and stark. As I entered, I saw at once the painting on the wall opposite the door. There were the rocks I had just been walking on, and the famous lighthouse, and a stretch of the night sea with God's hands reaching down, cupping an inscription:

> A Memorial to the 229 souls who perished in these waters of St. Margaret's Bay, Nova Scotia, September 2, 1998, at 10:30 pm ADT, while being aboard Swissair Flight 111. Embraced in God's Eternal Care.

The picture took my breath away. I felt the passionate emotions of people whose lives had been marked by picking up the pieces of Flight 111. The building resonated with their prayers. I was glad that I had asked to stop.

During the early years of the twentieth century, two other maritime disasters were visited upon Halifax. The second and worst occurred in the harbour, on 7 December, 1917 when a ship, loaded with explosives, was rammed by another. Two thousand people, many of them on shore, died in the ensuing explosion, the largest caused by man before the atomic age. This tragic story was new to me, but the first disaster, the loss of the *Titanic* on its maiden voyage in 1912, had always been part of my imaginative experience. It was the *Titanic* that I'd thought of when I first pictured Swissair 111 settling into the sea. Although that image was wrong, the two tragedies remained linked in my mind as disasters of hubris, of man's flawed belief that he can conquer nature, that he can bridge the Atlantic itself in days, or even hours.

My family was as interested in the *Titanic* as I was, so in Halifax's Maritime Museum the four of us pored over the *Titanic* exhibits. 'As far as the eye could see,' an eye-witness wrote, 'the ocean was strewn with wreckage and debris, with bodies bobbing up and down in the cold sea.' For a while, I lost my own sadness in contemplating the tragedy of the *Titanic*. I looked in wonder at the bits of wood that had been part of the great ship. And I disappeared into the personal history of those earlier victims, especially that of a man named Michel Navratil.

Separated from his wife, Navratil kidnapped their two children and fled aboard *Titanic* under an assumed name. He managed to get Edmund and Michel into the very last lifeboat. The boys were eventually reunited with their mother. Their father, however, was buried under his false name, Louis Hoffman, in the Baron de Hirsch Jewish Cemetery and not identified for several years.

Poor Michel! What was he thinking, as he entrusted his boys to the last lifeboat in those terrible scenes of chaos and fear? Did he wonder if it was a judgment on the thing he had done?

❧

1 September 1999. The hotel's breakfast room was busy. Two small boys caught my eye. Aged around six and nine, they were dressed beautifully in trousers, waistcoats and Mickey Mouse ties. Their eyes were bright, their faces intelligent. They were gorgeous, a moment's distraction from the difficult day ahead.

For the Swissair Memorial had begun. Most of the day was to be taken up with the dedication of two monuments on the Nova Scotia coast, the first at Bayswater, the second at Whalesback. The two monuments, together with the unmarked crash site at sea, had been conceived as a triangle of remembrance.

The day was hot and bright. A fleet of buses had been organised for the 700 mourners, and boarding was time-consuming. While we waited, I took out a wooden cross that had belonged to Frank's mother, who had died a couple of weeks earlier, a cross that I held throughout the morning. In time, the convoy set off, escorted by police outriders. People stood by the roadside as we passed, with ordinary traffic stopped. The momentousness of the occasion was affecting.

When we got to Bayswater, we were directed up a leafy pathway towards the monument. Along the way, hundreds of people stood to attention, Canadians who had taken part in the recovery. From tables heavy with flowers, I chose an iris. Patty had given me a dozen small iris-embellished gifts: postcards, wrapping paper, pictures. Now, for the first and last time, I had an iris for her.

At the top of the path, the memorial wall rose, covered with names. Its granite length was aligned to point towards the crash site. I stood before Patty's name in a trance of grief and realisation. Standing in that formal, studied place blasted away the shreds of unreality that had clung to her death. Here, with that massive structure in front of me, there was no ambiguity.

After a long time, I read some of the other names, shuddered at the repetitions, shuddered at the annihilation of couples, friends, entire families. In the centre of the plaza were three oversized coffins, burnished in bronze and gold, two above the ground, the third lowered already into a crypt. They held, I knew, a token quantity of unidentified remains: bone, splinters of bone, scraps of flesh. Remains too small, or too difficult, to identify. Only a year before, I thought, every bit and scrap had been vibrant, part of the miraculous human machine.

In the centre of the monument was the inscription:

IN MEMORY OF

THE 229 MEN, WOMEN AND CHILDREN

ABOARD SWISSAIR FLIGHT 111

WHO PERISHED OFF THESE SHORES

SEPTEMBER 2, 1998

THEY HAVE BEEN JOINED TO THE

SEA AND THE SKY

MAY THEY REST IN PEACE

This inscription, repeated in French – Canada's other official language – struck me as both poetic and literal. Patty, truly, had become part of the sea and the sky. Blood and flesh, in the sea. Smoke, in the sky. Ashes, in seawater and blown away in air.

There had been many buses behind us, and people were still flooding into the plaza, succumbing to emotion. The sorrow was all-encompassing. Suddenly in front of me, I saw the children from breakfast, the boys in their smart waistcoats and Mickey Mouse ties. The elder was poking disconsolately at the dust. The younger was concentrating on a toy soldier, clutched in his small hands. With them was an elegant woman in black and white, her eyes awash, her body trembling, and, also, an elderly couple. Appalled, I read the names on their security badges. I saw that the boys had lost their father, the woman her husband, and the elderly couple their son.

My own tears started again. Surely this was too much. Who could ever make up to those small boys the loss of their father?

to that woman, her partner? to that couple, their son? I had felt acutely the tragedy of Patty's leaving no children behind her. Now I saw that the alternative was as bad.

At the end of the service, the Lone Piper's Lament rang out into the hot, blue sky. As the crowded plaza cleared, I had no desire to leave. I went over again to the wall and touched Patty's name. The base of the wall was heaped with flowers. I buried my mother-in-law's cross among them. Reading the names adjacent to Patty's, I registered another Patricia. Patricia Ezell and Patricia Eberhart, strangers in life and neighbours in death.

As we finally left, I shook hands with some of the Canadians still lining the leafy path. Their good wishes and selfless efforts touched me, did something to make the unbearable bearable.

The buses wound through the heat along the spectacular shoreline of St. Margaret's Bay, through Aspotogan, Birchy Head, Fox Point, Queensland, Black Point, Upper Tantallon, French Village, Seabright, Indian Harbour, and, finally, Whalesback. The dedication at Bayswater had been for family members only. Whalesback was a public event, and thousands of people awaited our arrival. We followed a winding path through the sombre congregation, until stopped by a wall of people in front of us. When the ceremony began, I realised that we were within feet of the monument and a podium, which we could not see. So we listened to the words being spoken beside us, but amplified throughout the crowd, which was unsettling. When it was over, and the area cleared, we found the monument, touched it. Then, for a long time, the four of us sat together on the rocks, staring out to sea.

The second day of September, the first anniversary of Patty's death. With other mourners, we were taken by bus to the military base at Shearwater, where investigations into the cause of the crash were ongoing. The base sprawled like a giant's graveyard, with huge, tinny hangars heaving up from the earth. An air of desolation clung to them, vast surfaces to be scoured by storms, weighted down with ice.

From the beginning, I had longed to see the wreckage, as a way of getting close. I knew roughly what to expect, because the Transportation Safety Board of Canada had sent a video. But it is one thing to see pictures on a small screen, it is another to be surrounded by the physicality of the pulverised shards of a wide-bodied jet. So abundant were the fragments that it required two hangars and a yard to house them. The face of the chief investigator, Vic Gerden, was familiar from the video and newspapers. 'From the start,' he told us, 'it was clear that the plane's difficulties grew out of an electrical fire. But questions remain: what caused the fire? and how did it spread?'

Questions remain. What caused the fire? How did it spread?

In Hangar A, the fragments that interested them most were laid out on long tables. A young official handed me a piece of metal. 'From the skin of the aircraft,' he said. It was the size of my hand. Its edges were snapped and torn, curling up on one side. 'There are countless pieces like that,' the young man said. 'None of them big. Barring the engines and the landing gear, you'd be able to move every piece yourself.'

Fragments that I could move myself? I ran my finger along the jagged edge. Apparently, the plane had shattered like a light bulb.

We were shown ducting, piles of it, crushed and in pieces. Men and women were beating out the ducting, patching it together like an enormous jigsaw. By examining colour changes, they hoped to track the path of the fire. The cabin ceiling was burnt from the back of the cockpit, through the galley, into the first-class cabin, a distance of three to four metres. The investigators were reconstructing this area exactly. Fragments of wreckage dangled from a cockpit-shaped frame. But much was still missing. In a few days, we were told, they would be dredging up the seabed looking for more.

On the floor behind the cockpit-shaped frame, the interior of the cockpit was marked out. An investigator showed us the shredded remains of a seatbelt harness. He explained – not too graphically – how the seatbelt revealed that the co-pilot had been in his seat at the moment of impact. The captain, however, had not been wearing his seatbelt. 'Not necessarily because he wasn't in his seat,' the man added softly. 'He may have taken it off for mobility while tackling the plane's problems.' A picture of the dead pilot flickered in my mind. His name was Urs Zimmermann. Patty had introduced him to me once on a Swissair crew bus. I had seen footage of him on television from a Swissair training video. There had also been some funny home videos, shot on holiday with his family in Venice. 'When I want to be near him now,' his wife had said, 'I come to Halifax.' How I felt for her, Prisca Zimmermann, and her teenaged children, and the wife of the co-pilot, Sonja Loëw, who had three children under five when her husband died in the shredded harness.

I would have been glad of more time with the investigators, but new buses were arriving steadily, and we had to move on. Hangar J held the bulk of the wreckage in hundreds of cardboard cartons, each labelled with a seat number or other identifying information. The engines and landing gear dominated an outside yard. From the barrier, I stared at the grotesquely twisted engines, their blades snapped and warped. The landing gear, although retracted at impact, had snapped like a dry stick. The huge tyres hung in flitters.

A sense of desolation was everywhere.

Back at the hotel, we escaped for an hour to the sports facilities, avoiding a reception. Crowds deter me. I prefer quiet rooms, where I can study faces. I wasn't up to small talk, and my sons did not want to go at all. We arrived in time for the speakers: including the Canadian Prime Minister, the Premier of Nova Scotia, and the President of Switzerland. We listened for a while, then left.

The main event of the memorial had been timed to coincide with the last flight of the doomed airliner. It took place in the Citadel, a walled fortress perched on the heights of Halifax, a large space open to the skies and packed with thousands of people. It seemed as if we were reaching up towards the spirits of the dead. As the evening began with a contralto solo sung from the ramparts, I thought of Patty's plane, rolling away from its gate, joining the line of planes waiting to take off. I pictured her slight shoulders, her neat uniform. I saw her smile, as she chatted with passengers, stowed gear. Then I watched the plane accelerate down the runway, lifting off, all light and grace.

Silver bird soaring in the sky.

Patty, busy in the aisles. Passengers kicking off shoes, stretching back. Drinks, books, laptops. The ordinary about to become horrific.

The ceremony proceeded. People spoke, groups sang. Two young women, twins who had lost their father, read a poem written by one, first in English then in French. Tears came and went. An airplane tracked its way through the sky. Candles were lit, one for every soul. They were carried to the front. I watched Richard, candle in hand, his young back straight, and I thought of how very much Patty had loved him. Was she up there, in that infinite dome, watching him bear the symbol of her life?

The candles lined the foot of the stage, a sea of shimmering light. The memorial was not, I thought, about closure at all. How could there be closure, when lives had been changed for ever? But the ritual was drawing a line. It was acknowledging what had been too public a death. Now mourning, and the struggle to rebuild lives could continue privately.

When the ceremony was over, I began to talk with the young man who had been sitting next to me. His brother, Patrick, had died on the plane. Patrick had been in his twenties, the youngest of four sons. He had lived and worked in Manhattan, in the immediate area around Patty's apartment. Almost certainly, they had passed one another in the street. Patty and Patrick. His parents joined us, told me more. Patrick's trip to Geneva had been planned in his office. A gang had contemplated going, as many as eight at one stage. Then one by one, they had dropped out, leaving only a pair of adventurers, friends who had died together.

Patrick and Patty: strangers who had brushed shoulders in life and in death. Under the twinkling night sky, I listened. There were 229 unique stories, I realised, every one a narrative of joys and sorrows, each embroidered with the twists of fate that had led them to the same conclusion.

Epilogue

Tuesday, 25 August 1998. It is an excellent day. My eighteen-year-old son, Richard, receives in the mail his first-choice offer for college. He will be studying at Trinity College, Dublin. I telephone Mother to let her know. Then I telephone Patty.

'Why that's very good news,' she says, her New York twang zinging across the Atlantic. Patty adores Richard. He is her godson, named for our Dad. Whenever she can, Patty takes him places, loving it when people think that he's her son. She is thrilled he has done so well, ecstatic that he'll be going to Trinity College.

Then her mind jumps to my imminent visit to New York. Someone to do things with, if only for the week. 'I'm swapping a trip,' she says, 'so I'll be completely free.' She fires out a list of the restaurants and stores that we'll visit. 'Fabulous,' she says, 'you'll love them. And we'll do Connecticut before the opera. I'll be coming in on the 110 when your flight arrives. So we'll meet at the airport. Mom can come out on the bus, and we'll drive straight to Connecticut.'

As I listen to these permutations, I can see Patty frown across the ocean. I can see the telephone on her kitchen counter, and the red Italian biscuit tin on the shelf over her head. *Amaretti di Saronno* it says on the tin. 'No,' Patty sighs. 'I can't make up my mind. I'll let you know when I decide. There's so much to pack in, I need more hours in the day. Anyway, I've got to run. Bye-bye-ee!'

'Good-bye!'

And then, she is gone.

Acknowledgements

Many people contributed to the writing of *Blunt Trauma*. During the year following the last flight of Swissair 111, my son Richard Bannister faithfully gathered information from the Internet for me. At the same time, in London, Alexandra Erskine monitored the UK press. When in Halifax, I discovered Stephen Kimber's account, *Flight 111, The Tragedy of the Swissair Crash*, which provided some insight into the workings of the various Canadian emergency organisations, for which I am grateful.

While writing, many individuals offered criticism and encouragement. As I wrote the difficult second draft, Mary Hyde Hood read my daily output. Mary's sister, Sylvia Hyde Koester, and her mother-in-law, Mrs. Hood, with whom Patty once lived, also scrutinised that draft, and made valuable suggestions. This connection with the extended Hood and Hyde families, which grew out of my returning two Christmas cards addressed to Patty, seems to have been brought about by Patty, and has improved my life as well as this book. Additionally, I am indebted to Mary Rose Callaghan for the countless hours shared talking about bereavement, hours that helped with more than the writing. For two years, Mary Rose, Maïrïde, Eílís, Celia, Colette, Liz, Lia, Sheila, Maggie, Catherine, Carmel and Clairr all listened to parts of the manuscript, offering much helpful advice.

On a personal level, I would like to thank Avril and Carl Foran, Kevin Moran and the late Mona Moran, Mary, Harry and Niall Power, and Sarah Hayes, all who provided practical support when I most needed it. Equally, my debts to David and Franny Eberhart, Rick and Mary Lee Watson can never be repaid. Their unstinting kindnesses enabled me to keep my head above water.

I am grateful for the support of Delta Airlines. The letters received from hundreds of Patty's friends and colleagues meant a great deal, and have been preserved. I would like to single out Maritza Biscaino, Kent Jackson and John Walters for their personal attentions. Just so, I thank the countless Canadians who contributed to making our trip to Nova Scotia in the autumn of 1999 more bearable. I will never know most of their names, but their generosity has touched me forever. I acknowledge with awe the thousands of Canadians, volunteers and workers who faced courageously the tragedy that descended upon them on the evening of 2 September, 1998.

A particular thanks is due to my agent, Jonathan Williams, whose efforts and intelligence inform every page, and to John Davey and Susan Waine, who have made this book a reality. Finally, for Frank, Richard and Andrew, thank you for being the people you are.